UNITEXT for Physics

UNITEXT for Physics series publishes textbooks in physics and astronomy, characterized by a didactic style and comprehensiveness. The books are addressed to upper-undergraduate and graduate students, but also to scientists and researchers as important resources for their education, knowledge, and teaching.

More information about this series at https://link.springer.com/bookseries/13351

Philippe Jetzer

Applications of General Relativity

With Problems

 Springer

Philippe Jetzer
Department of Physics
University of Zurich
Zürich, Switzerland

ISSN 2198-7882   ISSN 2198-7890 (electronic)
UNITEXT for Physics
ISBN 978-3-030-95720-9   ISBN 978-3-030-95718-6 (eBook)
https://doi.org/10.1007/978-3-030-95718-6

This Springer imprint is published by the registered company Springer Nature Switzerland AG
The registered company address is: Gewerbestrasse 11, 6330 Cham, Switzerland

For Fiorenza

Preface

The aim of this book is to present in a comprehensive way several advanced topics related to general relativity. In Chap. 1, I review the basic knowledge on general relativity needed for the understanding of the next topics. The following chapter is dedicated to some applications of general relativity, among which it is useful to mention gravitational lensing, relativistic stellar structure, time delay as well as geodetic precession. Following, I discuss in detail gravitational waves and the post-Newtonian approximation. Chapter 4 is devoted to several topics related to black holes, including the treatment of spinors in curved spacetime and the four laws of black hole dynamics. Finally, the last chapter covers the equivalence principle and ways to test it. The selection of the topics included in this book has been made according to ongoing important research activities, both theoretically and experimentally, in order to provide advanced undergraduate and graduate students with a modern pedagogical text designed to broaden the knowledge acquired during a core course on general relativity. I developed this book from a series of lectures given on a yearly basis with the intention of offering a gentle introduction to the subject, which helps reading the more specialized literature and that can be used as a first reading to get quickly into the field when starting research. The book includes problems to practice understanding of key concepts as well as a list of selected references to deepen the material.

Zürich, Switzerland
December 2021

Philippe Jetzer

Acknowledgments

This book has grown up from lectures I gave in the last years for students of both the University of Zürich and ETH as a continuation of my lecture on general relativity. The notes, originally intended as *internal* lecture notes, have been typed by several students attending the courses as well as by some of my Ph.D. students. I thank them all for their invaluable contribution and the numerous suggestions, improvements and corrections, in alphabetic order: Raymond Angélil, Tobias Baldauf, Simone Bavera, Yannick Boetzel, Adrian Boitier, Felix Hähl, D. Moosbrugger and especially Michael Ebersold, who took care of the problems. I wish to thank several collaborators in the team of Springer, who contacted and encouraged me to write this book, in particular Lisa Scalone. After some initial reluctance in doing it, they convinced and helped me all along the different editorial processes. Without their patience the book would not have been written.

Contents

Chapter 1
Elements of General Relativity

In this chapter we give a brief introduction to general relativity (GR), and in particular we discuss the main equations we will need in the following ones. We keep this chapter very short, and for more detailed derivation and explanation of the various formulae and concepts we refer to the literature, as indeed there are many excellent books on general relativity, some of which we list at the end of this chapter for further reading, without, however, any pretension of being exhaustive [1–12].

1.1 Einstein's Field Equations

General relativity is based on the principle of equivalence of gravitation and inertia, which tells us how an arbitrary physical system responds to an external gravitational field. The *equivalence principle* is as formulated by Einstein:

1. Inertial and gravitational mass are equal.
2. Gravitational forces are equivalent to inertial forces.
3. In a local inertial frame, we experience the known laws of special relativity without gravitation.

The physical laws are relations among tensors (scalars and vectors being tensors of rank 0 and 1, respectively). Thus the physical laws read the same in all coordinate systems (provided the physical quantities are transformed suitably) and satisfy the *general covariance principle* (which means they have the same form).

> Covariance principle: Write the appropriate special relativistic equations that hold in the absence of gravitation, replace $\eta_{\alpha\beta}$ by $g_{\alpha\beta}$, and replace all derivatives with covariant derivatives $(, \rightarrow ;)$. The resulting equations will be generally covariant and true in the presence of gravitational fields.

© The Author(s), under exclusive license to Springer Nature Switzerland AG 2022
P. Jetzer, *Applications of General Relativity*, UNITEXT for Physics,
https://doi.org/10.1007/978-3-030-95718-6_1

The local coordinate expression for the covariant derivative for a tensor of rank (p, q) is given by:

$$T^{i_1 \ldots i_p}{}_{j_1 \ldots j_q; k} = T^{i_1 \ldots i_p}{}_{j_1 \ldots j_q, k} + \Gamma^{i_1}_{kl} T^{l i_2 \ldots i_p}{}_{j_1 \ldots j_q} + \cdots + \Gamma^{i_p}_{kl} T^{i_1 \ldots i_{p-1} l}{}_{j_1 \ldots j_q}$$
$$- \Gamma^l_{k j_1} T^{i_1 \ldots i_p}{}_{l j_2 \ldots j_q} - \cdots - \Gamma^l_{k j_q} T^{i_1 \ldots i_p}{}_{j_1 \ldots j_{q-1} l}. \qquad (1.1)$$

Examples:

- Contravariant and covariant vector fields:

$$\xi^i{}_{;k} = \xi^i{}_{,k} + \Gamma^i_{kl} \xi^l, \qquad (1.2)$$

$$\eta_{i;k} = \eta_{i,k} - \Gamma^l_{ki} \eta_l, \qquad (1.3)$$

- Kronecker tensor:

$$\delta^i{}_{j;k} = 0, \qquad (1.4)$$

- Tensor of rank $(1, 1)$:

$$T^i{}_{k;r} = T^i{}_{k,r} + \Gamma^i_{rl} T^l{}_k - \Gamma^l_{rk} T^i{}_l. \qquad (1.5)$$

The covariant derivative of a tensor is again a tensor. Consider the covariant derivative of the metric $g_{\mu\nu}$:

$$g_{\mu\nu;\lambda} = \frac{\partial g_{\mu\nu}}{\partial x^\lambda} - \Gamma^\rho_{\lambda\mu} g_{\rho\nu} - \Gamma^\rho_{\lambda\nu} g_{\rho\mu}. \qquad (1.6)$$

The expressions for the Christoffel symbols $\Gamma^\rho_{\lambda\mu}$ are given by

$$\Gamma^\sigma_{\lambda\mu} = \frac{g^{\sigma\nu}}{2} \left(\frac{\partial g_{\mu\nu}}{\partial x^\lambda} + \frac{\partial g_{\lambda\nu}}{\partial x^\mu} - \frac{\partial g_{\mu\lambda}}{\partial x^\nu} \right). \qquad (1.7)$$

Note that the Γ's are symmetric in the lower indices $\Gamma^\kappa_{\mu\nu} = \Gamma^\kappa_{\nu\mu}$. Inserting them into (1.6) leads to

$$\boxed{g_{\mu\nu;\lambda} = 0.} \qquad (1.8)$$

This is not surprising since $g_{\mu\nu;\lambda}$ vanishes in locally inertial coordinates, and being a tensor it is then zero in all systems. This is a consequence of the equivalence principle. Notice that the lowering or raising of indices is done with $g_{\mu\nu}$ or $g^{\mu\nu}$, respectively.

The field equations, which determine the metric $g_{\alpha\beta}$, cannot be derived by using the covariance principle, since there is no equivalent equation in a local inertial coordinate system and thus in special relativity. We, therefore, have to make some assumptions to get them, and their validity relies then upon the verification of their predictions with experiments.

Requirements

- The Newtonian limit is well confirmed through all observations, in which case we have the following equation: $\Delta\phi = 4\pi G\rho$, where ϕ is the classical Newtonian potential, G Newton's constant and ρ the matter density.

Thus a generalization should be as follows: $G_{\mu\nu} = \frac{8\pi G}{c^4} T_{\mu\nu}$, where $T_{\mu\nu}$ is the matter tensor, and $G_{\mu\nu}$ (Einstein tensor) has to satisfy the following requirements:

1. $G_{\mu\nu}$ is a tensor (as well as $T_{\mu\nu}$).
2. $G_{\mu\nu}$ contains only terms up to second derivatives of the metric. $G_{\mu\nu}$ is then a linear combination of terms which are either linear in the second derivative of the metric $g_{\mu\nu}$ or quadratic in the first derivative of $g_{\mu\nu}$.
3. Since $T_{\mu\nu}$ is symmetric, $G_{\mu\nu}$ has also to be symmetric ($G_{\mu\nu} = G_{\nu\mu}$) and due to the fact that $T_{\mu\nu}$ is covariantly conserved, i.e., $T^{\mu\nu}{}_{;\nu} = 0$, it follows that $G_{\mu\nu}$ must satisfy $G_{\mu\nu}{}^{;\nu} = 0$ as well.
4. For the non-relativistic limit (weak stationary field) one expects then the following relation to hold

$$\Delta g_{00} = \frac{8\pi G}{c^4} T_{00}, \tag{1.9}$$

with $T_{00} \approx \rho c^2$ (all other T_{ij} being small) and thus $G_{00} \simeq \Delta g_{00}$. From the Newtonian limit of the equation of motion for a particle (see (1.43)) one gets the following equation for the g_{00} component of the metric tensor: $g_{00} \approx 1 + \frac{2\phi}{c^2}$.

We use the common notation such that "," means normal derivation, whereas ";" is the covariant derivative. Conditions (1)–(4) determine $G_{\mu\nu}$ uniquely [13]. (1) and (2) imply that $G_{\mu\nu}$ has to be a linear combination

$$G_{\mu\nu} = aR_{\mu\nu} + bRg_{\mu\nu} \tag{1.10}$$

of $R_{\mu\nu}$, the Ricci tensor, and R, the Ricci scalar.[1]

The *Riemann tensor or curvature tensor* is defined as follows

$$R^i{}_{jkl} = \Gamma^i_{lj,k} - \Gamma^i_{kj,l} + \Gamma^s_{lj}\Gamma^i_{ks} - \Gamma^s_{kj}\Gamma^i_{ls}. \tag{1.11}$$

It is antisymmetric in the last two indices: $R^i{}_{jkl} = -R^i{}_{jlk}$.

The *Ricci tensor* is the following contraction of the curvature tensor:

$$\boxed{R_{jl} \equiv R^i{}_{jil} = \Gamma^i_{lj,i} - \Gamma^i_{ij,l} + \Gamma^s_{lj}\Gamma^i_{is} - \Gamma^s_{ij}\Gamma^i_{ls}.} \tag{1.12}$$

The *scalar curvature* is the trace of the Ricci tensor:

$$\boxed{R \equiv g^{lj} R_{jl} = R^l{}_l.} \tag{1.13}$$

---

[1] It can be shown that indeed the Ricci tensor is the only tensor made of the metric tensor $g_{\mu\nu}$ with first and second derivatives of it, which is linear in the second derivative [1].

The symmetry of $G_{\mu\nu}$ is automatically satisfied. The contracted Bianchi identity, such that $G_{\mu\nu}{}^{;\nu} = 0$, requires that $b = -\frac{a}{2}$. Thus we find

$$G_{\mu\nu} = a\left(R_{\mu\nu} - \frac{1}{2}g_{\mu\nu}R\right) = \frac{8\pi G}{c^4}T_{\mu\nu}. \qquad (1.14)$$

The constant a has to be determined by performing the Newtonian limit. Consider weak fields: $g_{\mu\nu} = \eta_{\mu\nu} + h_{\mu\nu}$, $|h_{\mu\nu}| \ll 1$ (non-relativistic velocities: $v^i \ll c$), then $|T_{ik}| \ll |T_{00}| \Rightarrow |G_{ik}| \ll |G_{00}|$. Computing the trace of $G_{\mu\nu}$ leads to:

$$g^{\mu\nu}G_{\mu\nu}\begin{cases} = a(R - 2R) = -aR & \text{from (1.14)} \\ \approx G_{00} = a\left(R_{00} - \frac{R}{2}\underbrace{g_{00}}_{\substack{\approx\eta_{00} \\ =1}}\right) = a(R_{00} - R/2). \end{cases}$$

Comparing the two results gives $R \approx -2R_{00}$, thus

$$G_{00} \simeq a\left(R_{00} - \frac{R}{2}\right) \simeq 2aR_{00}. \qquad (1.15)$$

For weak fields all terms quadratic in $h_{\mu\nu}$ can be neglected in the Riemann tensor; we get to leading order:

$$R_{\mu\nu} = R^\rho{}_{\mu\rho\nu} \simeq \frac{\partial\Gamma^\rho_{\mu\nu}}{\partial x^\nu} - \frac{\partial\Gamma^\rho_{\mu\nu}}{\partial x^\rho} \qquad (|h_{\mu\nu}| \ll 1). \qquad (1.16)$$

For weak stationary fields we find:

$$R_{00} = -\frac{\partial\Gamma^i_{00}}{\partial x^i} \quad \text{with} \quad \Gamma^i_{00} = \frac{1}{2}\frac{\partial g_{00}}{\partial x^i}. \qquad (1.17)$$

Thus $G_{00} \approx -2a\partial\Gamma^i_{00}x^i = -a\Delta g_{00} \overset{!}{=} \Delta g_{00}$, therefore $a = -1$. Einstein's field equations are (derived in 1915 by Albert Einstein):

$$\boxed{R_{\mu\nu} - \frac{R}{2}g_{\mu\nu} = -\frac{8\pi G}{c^4}T_{\mu\nu}.} \qquad (1.18)$$

Together with the geodesic equation (see (1.43)), these are the fundamental equations of general relativity. By contraction of (1.18), we find also

$$R^\mu{}_\mu - \frac{R}{2}\underbrace{\delta^\mu{}_\mu}_{=4} = -R = -\frac{8\pi G}{c^4}T. \qquad (1.19)$$

R can be expressed in (1.18) in terms of $T = T^{\mu}{}_{\mu}$, and we get:

$$R_{\mu\nu} = -\frac{8\pi G}{c^4}\left(T_{\mu\nu} - \frac{T}{2}g_{\mu\nu}\right) \tag{1.20}$$

an equivalent version of the field equations. For the vacuum case, where $T_{\mu\nu} = 0$, we have

$$R_{\mu\nu} = 0. \tag{1.21}$$

Einstein's equations constitute a set of nonlinear coupled partial differential equations whose general solution is not known. Since the Ricci tensor is symmetric, the Einstein equations are a set of 10 algebraically independent second-order differential equations for $g_{\mu\nu}$.

The Einstein equations are generally covariant, so that they can at best determine the metric up to coordinate transformations ($\rightarrow$4 functions). Therefore we expect only 6 independent generally covariant equations for the metric. Indeed the (contracted) Bianchi identities tell us that $G^{\nu}{}_{\mu;\nu} = 0$, and hence there are 4 differential relations among Einstein's equations. Bianchi identities can also be seen as a consequence of the general covariance of the Einstein equations.

1.1.1 The Cosmological Constant

As a generalization, one can relax condition (2) and add a linear term in $g_{\mu\nu}$. The field equations become

$$R_{\mu\nu} - \frac{R}{2}g_{\mu\nu} + \Lambda g_{\mu\nu} = -\frac{8\pi G}{c^4}T_{\mu\nu}, \tag{1.22}$$

where Λ is a constant: the cosmological constant (in our units its dimension is: $[\Lambda] = L^{-2}$, L being a length). For point (4) the Newtonian limit of (1.22) leads to

$$\Delta\phi = 4\pi\rho G + \frac{c^2}{2}\Lambda. \tag{1.23}$$

The right-hand side can also be written as $4\pi G(\rho + \rho_{\text{vac}})$, with

$$c^2\rho_{\text{vac}} = \frac{c^4}{8\pi G}\Lambda. \tag{1.24}$$

Λ corresponds to the (constant) energy density of empty space (vacuum). $\Lambda^{-1/2}$ (its unit is a length) has to be much larger than the dimension of the solar system in order to satisfy the very precise constraints coming from the ephemeris of the planets in the solar system.

1.2 Static Isotropic Metric

1.2.1 Form of the Metric

For the gravity field of celestial bodies like Earth and Sun we can assume a spherically symmetric distribution of the matter (with very small rotation velocities $v^i \ll c$). Thus we need a spherically symmetric and static solution for the metric $g_{\mu\nu}(x)$. We first give the general form of such a metric (static and isotropic), which we then use as an ansatz to solve the field equations. For $r \to \infty$, the Newtonian gravitational potential $\phi = -\frac{GM}{r}$ goes to zero. Thus, asymptotically, the metric should be Minkowskian: $\mathrm{d}s^2 \underset{r \to \infty}{=} c^2\mathrm{d}t^2 - \mathrm{d}r^2 - r^2(\mathrm{d}\theta^2 + \sin^2\theta\mathrm{d}\varphi^2)$, in spherical coordinates r, θ, φ and t. Thus,

$$\mathrm{d}s^2 = B(r)c^2\mathrm{d}t^2 - A(r)\mathrm{d}r^2 - C(r)r^2(\mathrm{d}\theta^2 + \sin^2\theta\mathrm{d}\varphi^2). \tag{1.25}$$

Due to isotropy and time independence, A, B and C cannot depend on θ, φ and t (and no linear terms in $\mathrm{d}\theta$ and $\mathrm{d}\varphi$). Freedom in the choice of coordinates allows to introduce a new radial coordinate in (1.25): $C(r)r^2 \to r^2$, thus $C(r)$ can be absorbed into r. We get the standard form:

$$\mathrm{d}s^2 = B(r)c^2\mathrm{d}t^2 - A(r)\mathrm{d}r^2 - r^2(\mathrm{d}\theta^2 + \sin^2\theta\mathrm{d}\varphi^2) \tag{1.26}$$

(θ and φ have the same significance as in Minkowski coordinates). Due to our asymptotic requirements ($r \to \infty$) we can assume that $B(r) \to 1$ and $A(r) \to 1$.

1.2.2 Christoffel Symbols and Ricci Tensor for the Spherically Symmetric Case

For a spherically symmetric case, assuming, e.g., that the mass distribution is spherically symmetric and not rotating, the Einstein field equations get simpler. The metric tensor $g_{\mu\nu}$ is then diagonal and its components can be written as follows

$$g_{00} = B(r) \qquad g_{11} = -A(r) \qquad g_{22} = -r^2 \qquad g_{33} = -r^2\sin^2\theta \tag{1.27}$$

$$g^{00} = \frac{1}{B(r)} \qquad g^{11} = -\frac{1}{A(r)} \qquad g^{22} = -\frac{1}{r^2} \qquad g^{33} = -\frac{1}{r^2\sin^2\theta}. \tag{1.28}$$

In this case the nonvanishing components of the Christoffel symbols are

$$\Gamma^0_{01} = \Gamma^0_{10} = \frac{B'}{2B} \qquad\qquad \Gamma^1_{00} = \frac{B'}{2A} \qquad \Gamma^1_{11} = \frac{A'}{2A}$$

$$\Gamma^2_{12} = \Gamma^2_{21} = \frac{1}{r} \qquad\qquad \Gamma^1_{22} = -\frac{r}{A} \qquad \Gamma^1_{33} = -\frac{r^2 \sin^2 \theta}{A} \qquad (1.29)$$

$$\Gamma^3_{13} = \Gamma^3_{31} = \frac{1}{r} \qquad\qquad \Gamma^3_{23} = \Gamma^3_{32} = \cot \theta \qquad \Gamma^2_{33} = -\sin \theta \cos \theta,$$

where $'$ stands for $\frac{\partial}{\partial r}$. With

$$-g = r^4 AB \sin^2 \theta, \qquad (1.30)$$

we get

$$\left(\Gamma^\rho_{\mu\rho} \right) = \left(\partial \ln \sqrt{-g} x^\mu \right) = \left(0, \frac{2}{r} + \frac{A'}{2A} + \frac{B'}{2B}, \cot \theta, 0 \right). \qquad (1.31)$$

For the Ricci tensor as given in Eq. (1.12) we thus get

$$R_{00} = -\frac{B''}{2A} + \frac{A'B'}{2A^2} + \frac{B'^2}{2AB} - \frac{B'}{2A} \left(\frac{2}{r} + \frac{A'}{2A} + \frac{B'}{2B} \right),$$

$$= -\frac{B''}{2A} + \frac{B'}{4A} \left(\frac{A'}{A} + \frac{B'}{B} \right) - \frac{B'}{rA}, \qquad (1.32)$$

$$R_{11} = +\frac{B''}{2B} - \frac{B'}{4B} \left(\frac{A'}{A} + \frac{B'}{B} \right) - \frac{A'}{rA}, \qquad (1.33)$$

$$R_{22} = -1 - \frac{r}{2A} \left(\frac{A'}{A} - \frac{B'}{B} \right) + \frac{1}{A}, \qquad (1.34)$$

$$R_{33} = R_{22} \sin^2 \theta. \qquad (1.35)$$

The non-diagonal components of $R_{\mu\nu}$ with $\mu \neq \nu$ vanish.

1.2.3 Robertson Expansion

Even without knowing the solution to the field equations, we can give an expansion of the metric for weak fields outside the mass distribution, assumed to be spherically symmetric and non-rotating. The metric can only depend on the total mass of the considered object (Earth or Sun for instance), on the distance from it and on the constants G and c. Since A and B are dimensionless, they can only depend on a combination of the dimensionless quantity $\frac{GM}{c^2 r}$. For $\frac{GM}{c^2 r} \ll 1$ we can then have the following expansion:

$$B(r) = 1 - 2\frac{GM}{c^2 r} + 2(\beta - \gamma)\left(\frac{GM}{c^2 r}\right)^2 + \cdots$$

$$A(r) = 1 + 2\gamma \frac{GM}{c^2 r} + \cdots \tag{1.36}$$

which is the *Robertson expansion*. The linear term in $B(r)$ has no free parameter since it is constrained by the Newtonian limit: $g_{00} \simeq 1 + 2\frac{\phi}{c^2}$, $\phi = -\frac{GM}{r}$ (Newtonian potential), therefore $B \to g_{00}$. The coefficient $2(\beta - \gamma)$ is defined this way for "historical reasons," and β and γ are independent coefficients. In the solar system, $\frac{GM}{c^2 r} \leq \frac{GM}{c^2 R_\odot} \simeq 2 \times 10^{-6}$, and thus only linear terms in γ and β play a role. For general relativity: $\gamma = \beta = 1$ (Newtonian gravity: $\gamma = \beta = 0$).

1.2.4 Schwarzschild Metric

We assume a static, spherically symmetric mass distribution with finite extension:

$$\rho(r) \begin{cases} \neq 0 & r \leq r_0 \\ = 0 & r > r_0. \end{cases}$$

Similarly, the pressure $P(r)$ vanishes for $r > r_0$. The four velocity vector within the mass distribution in the static case is $u^\mu = (u^0 = \text{constant}, 0, 0, 0)$. This way, the energy-momentum tensor (describing matter) does not depend on time. We then adopt the ansatz for the metric elaborated in (1.26): $g_{\mu\nu} = \text{diag}(B(r), -A(r), -r^2, -r^2 \sin^2 \theta)$. Outside the mass distribution ($r \geq r_0$), the Ricci tensor vanishes: $R_{\mu\nu} = 0$. We have already calculated the coefficients $R_{\mu\nu}$ in Eqs. (1.32)–(1.35). For $\mu \neq \nu$, $R_{\mu\nu} = 0$ is trivially satisfied, while the diagonal components have to be set to zero: $R_{00} = R_{11} = R_{22} = R_{33} = 0$ ($r \geq r_0$).

Consider $\dfrac{R_{00}}{B} + \dfrac{R_{11}}{A} = -\dfrac{1}{rA}\left(\dfrac{B'}{B} + \dfrac{A'}{A}\right) = 0$ and thus $\dfrac{\mathrm{d}}{\mathrm{d}r}(\ln AB) = 0$ (since $rA \neq 0$) or $AB = \text{constant}$. For $r \to \infty$ we require $A = B = 1$ (indeed, at infinity we assume to recover Minkowski metric), therefore $AB = 1 \Rightarrow A(r) = \frac{1}{B(r)}$. Introducing this into R_{22} (1.34) and R_{11} (1.33) leads to

$$R_{22} = -1 + rB' + B = 0, \tag{1.37}$$

$$R_{11} = \frac{B''}{2B} + \frac{B'}{rB} = \frac{rB'' + 2B'}{2rB} = \frac{1}{2rB}\frac{\mathrm{d}R_{22}}{\mathrm{d}r} = 0. \tag{1.38}$$

With (1.37), (1.38) is automatically satisfied (since $R_{22} = 0$ also its derivative vanishes). We write (1.37) as

$$\frac{\mathrm{d}}{\mathrm{d}r}(rB) = 1. \tag{1.39}$$

We integrate it and get $rB = r + \underbrace{\text{constant}}_{-2a} = r - 2a$. Then

$$B(r) = 1 - \frac{2a}{r},$$
$$A(r) = \frac{1}{1 - \frac{2a}{r}}, \qquad (1.40)$$

for $r \geq r_0$. This solution for the vacuum Einstein field equations was found in 1916 by Schwarzschild. The *Schwarzschild metric* is[2]

$$\boxed{ds^2 = \left(1 - \frac{2a}{r}\right) c^2 dt^2 - \frac{dr^2}{1 - \frac{2a}{r}} - r^2(d\theta^2 + \sin^2\theta\, d\varphi^2).} \qquad (1.41)$$

The constant can be determined by considering the Newtonian limit:

$$g_{00} = B(r) \xrightarrow[r\to\infty]{} 1 + 2\frac{\phi}{c^2} = 1 - 2\frac{GM}{c^2 r} = 1 - \frac{2a}{r}.$$

Thus one introduces the so-called *Schwarzschild radius*:

$$\boxed{r_S = 2a = \frac{2GM}{c^2}.} \qquad (1.42)$$

The Schwarzschild radius of the Sun is $r_{s,\odot} = \frac{2GM_\odot}{c^2} \simeq 3$ km ($M_\odot \simeq 2 \times 10^{30}$ kg, $R_\odot = 7 \times 10^5$ km) thus $\frac{r_{s,\odot}}{R_\odot} = \frac{2GM_\odot}{c^2 R_\odot} \simeq 4 \times 10^{-6}$. A clock at rest in r has the proper time $d\tau = \sqrt{B}\, dt$; thus $\frac{dt}{d\tau}$ diverges at $r \to r_S$. This implies that a photon emitted at $r = r_S$ will be infinitely redshifted (t is not a good coordinate either for events taking place at $r \leq r_S$). A star, whose radius r_{star} is smaller than r_S, is a black hole since photons emitted at its surface cannot reach regions with $r > r_S$.

Expanding the Schwarzschild metric in power of $\frac{r}{r_S}$ and comparing it with the Robertson expansion (1.36), one finds $\beta = \gamma = 1$ as required for general relativity.

1.3 General Equations of Motion

We now consider the motion of a freely falling material particle or photon in a static isotropic gravitational field (e.g., motion of planets around the Sun). For the relativistic orbit $x^k(\lambda)$ of a particle in a gravitational field we have:

---

[2] Apparently it seems that the Schwarzschild metric is singular for $r = r_s$, but this is not the case. It can be shown that it is only an artifact of the coordinate choice.

$$\frac{d^2 x^k}{d\lambda^2} = -\Gamma^k_{\mu\nu} \frac{dx^\mu}{d\lambda} \frac{dx^\nu}{d\lambda} \tag{1.43}$$

and

$$g_{\mu\nu} \frac{dx^\mu}{d\lambda} \frac{dx^\nu}{d\lambda} = \left(\frac{ds}{d\lambda}\right)^2 = c^2 \left(\frac{d\tau}{d\lambda}\right)^2 = \begin{cases} c^2 & m \neq 0, \quad \lambda = \tau \\ 0 & m = 0 \end{cases}. \tag{1.44}$$

For a massive particle we can take the proper time as a parameter for the trajectory or orbit ($d\lambda = d\tau$). For massless particles one has to choose another parameter. For the spherically symmetric gravitational field, we use the metric ($r > r_\odot$, radius of the star)

$$ds^2 = B(r)c^2 dt^2 - dr^2 A(r) - r^2(d\theta^2 + \sin^2\theta d\varphi^2), \tag{1.45}$$

with the coordinates $(x^0, x^1, x^2, x^3) = (ct, r, \theta, \varphi)$. Equations (1.43)–(1.45) define the relativistic Kepler problem. Using the Christoffel symbols given in (1.29), we get for (1.43):

$$\frac{d^2 x^0}{d\lambda^2} = -\frac{B'}{B} \frac{dx^0}{d\lambda} \frac{dr}{d\lambda}, \tag{1.46}$$

$$\frac{d^2 r}{d\lambda^2} = -\frac{B'}{2A} \left(\frac{dx^0}{d\lambda}\right)^2 - \frac{A'}{2A} \left(\frac{dr}{d\lambda}\right)^2 + \frac{r}{A} \left(\frac{d\theta}{d\lambda}\right)^2 + \frac{r^2 \sin^2\theta}{A} \left(\frac{d\varphi}{d\lambda}\right)^2, \tag{1.47}$$

$$\frac{d^2\theta}{d\lambda^2} = -\frac{2}{r} \frac{d\theta}{d\lambda} \frac{dr}{d\lambda} + \sin\theta \cos\theta \left(\frac{d\varphi}{d\lambda}\right)^2, \tag{1.48}$$

$$\frac{d^2\varphi}{d\lambda^2} = -\frac{2}{r} \frac{d\varphi}{d\lambda} \frac{dr}{d\lambda} - 2\cot\theta \frac{d\theta}{d\lambda} \frac{d\varphi}{d\lambda}. \tag{1.49}$$

Equation (1.48) can be solved by

$$\theta = \frac{\pi}{2} = \text{constant}. \tag{1.50}$$

Indeed, without loss of generality we can choose the coordinate system such that $\theta = \frac{\pi}{2}$, and this way the trajectory lies on the plane with $\theta = \frac{\pi}{2}$. The condition $\frac{d^2\theta}{d\lambda^2} = 0$ corresponds to angular momentum conservation. With (1.50) we get for (1.49):

$$\frac{1}{r^2} \frac{d}{d\lambda} \left(r^2 \frac{d\varphi}{d\lambda}\right) = 0, \tag{1.51}$$

which leads to

$$r^2 \frac{d\varphi}{d\lambda} = \text{constant} = l. \tag{1.52}$$

l can be interpreted as the (orbital) angular momentum (per unit mass). Equations (1.50) and (1.52) follow from angular momentum conservation, which is a consequence of spherical symmetry (rotation invariance).

Equation (1.46) can be written as ($B = B(r(\lambda))$)

$$\frac{\mathrm{d}}{\mathrm{d}\lambda}\left(\ln\left(\frac{\mathrm{d}x^0}{\mathrm{d}\lambda}\right) + \ln B\right) = 0, \tag{1.53}$$

which can be integrated as $\ln\left[\left(\frac{\mathrm{d}x^0}{\mathrm{d}\lambda}\right)B\right] = \text{constant}$ or

$$B\frac{\mathrm{d}x^0}{\mathrm{d}\lambda} = \text{constant} = F. \tag{1.54}$$

In (1.47) we use (1.50), (1.52) and (1.54) and get:

$$\frac{\mathrm{d}^2 r}{\mathrm{d}\lambda^2} + \frac{F^2 B'}{2AB^2} + \frac{A'}{2A}\left(\frac{\mathrm{d}r}{\mathrm{d}\lambda}\right)^2 - \frac{l^2}{Ar^3} = 0. \tag{1.55}$$

We multiply it with $2A\left(\frac{\mathrm{d}r}{\mathrm{d}\lambda}\right)$ and get[3]

$$\frac{\mathrm{d}}{\mathrm{d}\lambda}\left[A\left(\frac{\mathrm{d}r}{\mathrm{d}\lambda}\right)^2 + \frac{l^2}{r^2} - \frac{F^2}{A}\right] = 0. \tag{1.56}$$

Integration gives

$$A\left(\frac{\mathrm{d}r}{\mathrm{d}\lambda}\right)^2 + \frac{l^2}{r^2} - \frac{F^2}{B} = -\epsilon = \text{constant}. \tag{1.57}$$

Integrating it once more we get $r = r(\lambda)$. Inserting then this result into (1.52) and (1.54), we obtain with one more integration $\varphi = \varphi(\lambda)$ and $t = t(\lambda)$. Next we eliminate λ and get $r = r(t)$ and $\varphi = \varphi(t)$. Together with $\theta = \frac{\pi}{2}$, this is then a complete solution (generally it has to be done numerically).

Equation (1.44) becomes

$$g_{\mu\nu}\frac{\mathrm{d}x^\mu}{\mathrm{d}\lambda}\frac{\mathrm{d}x^\nu}{\mathrm{d}\lambda} = B\left(\frac{\mathrm{d}x^0}{\mathrm{d}\lambda}\right)^2 - A\left(\frac{\mathrm{d}r}{\mathrm{d}\lambda}\right)^2 - r^2\left(\frac{\mathrm{d}\theta}{\mathrm{d}\lambda}\right)^2 - r^2\sin^2\theta\left(\frac{\mathrm{d}\varphi}{\mathrm{d}\lambda}\right)^2 = \epsilon, \tag{1.58}$$

---

[3] Notice: $\dfrac{\mathrm{d}A}{\mathrm{d}\lambda} = A'\dfrac{\mathrm{d}r}{\mathrm{d}\lambda}$.

using (1.50), (1.52), (1.54) and (1.57). On the other hand

$$\epsilon = \begin{cases} c^2 & (m \neq 0) \\ 0 & (m = 0) \end{cases}.$$

We are left with two integration constants, F and l.

In the following a non-exhaustive list of reference books on general relativity is given.

References

1. S. Weinberg, *Gravitation and Cosmology* (Wiley, 1972)
2. C. Misner, K. Thorne, J. Wheeler, *Gravitation* (Freeman, 1973)
3. N. Straumann, *General Relativity with Applications to Astrophysics* (Springer Verlag, 2004)
4. R. Wald, *General Relativity* (Chicago University Press, 1984)
5. B. Schutz, *A First Course in General Relativity* (Cambridge, 1985)
6. R. Sachs, H. Wu, *General Relativity for Mathematicians* (Springer Verlag, 1977)
7. J. Hartle, *Gravity, An Introduction to Einstein's General Relativity* (Addison Wesley, 2002)
8. H. Stephani, *General Relativity* (Cambridge University Press, 1990)
9. S. Carroll, *Spacetime and Geometry: An Introduction to General Relativity* (Cambridge University Press, 2019)
10. T. Fliessbach, *Allgemeine Relativitätstheorie* (Spektrum Verlag, 1995)
11. M. Maggiore, *Gravitational Waves: Volume 1: Theory and Experiments* (Oxford University Press, 2007)
12. M. Maggiore, *Gravitational Waves: Volume 2: Astrophysics and Cosmology* (Oxford University Press, 2018)
13. D. Lovelock, J. math. Phys. **12**, 498–501 (1971)

Chapter 2
Some Applications of General Relativity

In this chapter we discuss some important applications of general relativity, in particular gravitational lensing, which is nowadays an important tool for the study of the universe: ranging from discovering planets in our galaxy to the distribution of dark matter at several scales and the determination of cosmological parameters. Moreover, we derive the relativistic stellar structure equations, which are relevant for the description of neutron stars. Other applications we mention are the geodetic precession and the Lense–Thirring effect. Both have been observed using satellites in Earth orbit. For the treatment of the Lense–Thirring effect we need first to discuss the linearized field equations, which will be of used also in the following chapters.

2.1 Gravitational Lensing

2.1.1 Historical Introduction

Even before the publication of Einstein's theory of general relativity, it was argued that gravity might influence the trajectories of light. Already Newton, in the first edition of his book on optics of 1704, discussed the possibility that celestial bodies could deflect the trajectory of light.

In 1804, the astronomer Soldner published an article in the *Berliner Astronomisches Jahrbuch* in which he computed the error induced by the light deflection on the determination of the position of stars [1]. He used to that purpose the Newtonian theory of gravity, assuming that the light consists of massive particles. He also estimated that a light ray which just grazes the surface of the sun would be deflected by a value of only 0.84 arc seconds. Within general relativity, this value is twice as much: 1.75 arc seconds. The first measurement of this effect has been made by British

© The Author(s), under exclusive license to Springer Nature Switzerland AG 2022
P. Jetzer, *Applications of General Relativity*, UNITEXT for Physics,
https://doi.org/10.1007/978-3-030-95718-6_2

astronomers during the total solar eclipse of May 29, 1919, with two expeditions, one at Sobral in northern Brazil and another on the island of Principe in the western coast of Africa. Both groups confirmed the value predicted by general relativity.

In 1936, Einstein published a short paper in Science in which he computed the light deflection of light coming from a distant star by the gravitational field of another star [2]. In the following year 1937, the swiss astronomer Zwicky wrote two short articles in Physical Review suggesting to consider galaxies as sources and lenses rather than stars as mentioned by Einstein [3, 4]. Moreover, he gave also a list of possible applications among which the possibility to better determine the total mass of galaxies, including their dark matter content. The first gravitational lens has been discovered in 1979, when spectra were obtained of two point-like quasars which lie only about 6 arc seconds away. The spectra showed that both objects have the same redshift and are thus at the same distance. Later on, the galaxy acting as lens for these two point-like quasars was found, making it clear that the two objects were actually the images of the same lensed quasar. In 1993 the first galactic microlensing events were observed, in which the sources are stars in the Large Magellanic Cloud. Later also many microlensing events were found in our galaxy by targeting as sources stars in the galactic bulge. This way also many planetary systems in our galaxy have been found.

In the last 40 years gravitational lensing has grown into a major area of research, which allows to probe the distribution of matter in galaxies and in galaxy clusters independently of the nature of the matter and thus to get information on the distribution of dark matter, but also to get independent measurements of the Hubble constant and other cosmological parameters [5, 6]. One can thus say that lensing has now become an essential tool for studying astrophysics and cosmology.

2.1.2   Point-Like Lens

The geometry of an homogeneous and isotropic expanding universe is described by the Friedmann–Lemaître–Robertson–Walker metric. All the inhomogeneities in the metric can be considered as local perturbations. Thus the trajectory of light coming from a distant source can be divided into three distinct pieces. In the first one the light coming from a distant source propagates in a unperturbed spacetime. Near the lens the trajectory gets modified due to the gravitational potential of the lens. In the third piece the light again travels in an unperturbed spacetime until it gets to the observer.

To first-order approximation, the region around the lens can be described by a flat Minkowskian spacetime with small perturbations induced by the gravitational potential of the lens according to the post-Newtonian approximation (see Sect. 3.4). This approximation is valid as long as the Newtonian potential Φ is small, which means $|\Phi| \ll c^2$, and if the peculiar velocity v of the lens is negligible compared to c. With these assumptions we can describe the light propagation near the lens in a flat spacetime.

The effect of the spacetime curvature on the light trajectory can be described as an effective refraction index, given by

$$n = 1 - \frac{2\Phi}{c^2} = \left(1 + \frac{2|\Phi|}{c^2}\right). \tag{2.1}$$

The Newtonian potential Φ is negative and vanishes asymptotically. As in geometrical optics a refraction index $n > 1$ means that the light travels with a speed which is less compared with its speed in the vacuum. Thus the effective speed of light in a gravitational field is given by

$$v = \frac{c}{n} \simeq c - \frac{2|\Phi|}{c}. \tag{2.2}$$

Since the effective speed of light is less in a gravitational field, the travel time gets longer compared to the propagation in empty space. The total time delay Δt, known as Shapiro delay (see Sect. 2.3), is obtained by integrating the second term (the one describing the delay) proportional to $|\Phi|$ in

$$dt = \frac{dx}{v} \simeq \frac{dx}{c}\left(1 + \frac{2|\Phi|}{c^2}\right) \tag{2.3}$$

along the light trajectory from the source x_S to the observer position x_O. This leads to

$$\Delta t = \int_{x_S}^{x_O} \frac{2|\Phi|}{c^3} dx. \tag{2.4}$$

The light trajectory is proportional to $\int n(\vec{x})d\vec{x}$, and the actual light path is the one which minimizes this functional, which corresponds to Fermat's principle, which states that: *The actual path between two points taken by a beam of light is the one which is traversed in the least time.* In other words the light path is the solution of the following variational problem

$$\delta \int n(\vec{x})d\vec{x} = 0 \Leftrightarrow \delta \int n(\vec{x}(\lambda))\left|\frac{d\vec{x}}{d\lambda}\right|d\lambda = 0. \tag{2.5}$$

Defining $L(\vec{x}, \dot{\vec{x}}) \equiv n(\vec{x})\sqrt{\dot{\vec{x}}^2}$ and $\dot{\vec{x}} \equiv \frac{d\vec{x}}{d\lambda}$ this problem leads to the Euler–Lagrange equation

$$\frac{d}{d\lambda}\frac{\partial L}{\partial \dot{\vec{x}}} - \frac{\partial L}{\partial \vec{x}} = 0, \tag{2.6}$$

with $\frac{\partial L}{\partial \vec{x}} = \frac{\partial n}{\partial \vec{x}} = \nabla n$, $\frac{\partial L}{\partial \dot{\vec{x}}} = n\dot{\vec{x}}$. By defining the tangent vector $\vec{e} := \dot{\vec{x}}$, assumed to be normalized by a suitable choice of the parameter λ, we get from the Euler–Lagrange equation

$$\frac{\mathrm{d}}{\mathrm{d}\lambda}(n\dot{\vec{x}}) - \nabla n = 0 \Leftrightarrow \frac{\mathrm{d}}{\mathrm{d}\lambda}(n\vec{e}) - \nabla n = n\dot{\vec{e}} + \vec{e}(\nabla n\vec{e}) - \nabla n = 0 \qquad (2.7)$$

$$\Leftrightarrow n\dot{\vec{e}} = \nabla n - \vec{e}(\vec{e} \cdot \nabla n) = \nabla_\perp n, \qquad (2.8)$$

where we used that $\nabla_\parallel = \vec{e}(\vec{e} \cdot \nabla)$ is the gradient component along the light path. Integrating this equation, using $\vec{e} \equiv \dot{\vec{x}}$ and $\nabla_\perp n = -\frac{2}{c^2}\nabla_\perp \Phi$ we get

$$\vec{\alpha} = \vec{e}_{\text{source}} - \vec{e}_{\text{observer}} = -\int_{\lambda_{\text{source}}}^{\lambda_{\text{obs}}} \frac{\mathrm{d}\vec{e}}{\mathrm{d}\lambda}\mathrm{d}\lambda = \frac{2}{c^2}\int_{z_{\text{source}}}^{z_{\text{obs}}} \nabla_\perp \Phi \mathrm{d}z. \qquad (2.9)$$

This holds for a light ray coming from the $-\vec{e}_z$ direction and passing by a point-like lens at $z = 0$ and $\vec{\alpha} = \vec{e}_{\text{source}} - \vec{e}_{\text{observer}}$ is the deflection angle for the light rays which go through a gravitational field. It is given by the integration of the gradient component of n (Eq. 2.1) perpendicular to the trajectory itself. For all astrophysical applications of interest the deflection angle is always extremely small, so that the computation can be substantially simplified by integrating $\nabla_\perp \Phi$ along an unperturbed path.

As an example we consider the deflection angle of a point-like lens of mass M. Its Newtonian potential is given by

$$\Phi(b, z) = -\frac{GM}{(b^2 + z^2)^{\frac{1}{2}}}, \qquad (2.10)$$

where b is the impact parameter of the unperturbed light ray and z denotes the position along the unperturbed path as measured from the point of minimal distance from the lens. This way we get

$$\nabla_\perp \Phi(b, z) = \frac{GM\vec{b}}{(b^2 + z^2)^{\frac{3}{2}}}, \qquad (2.11)$$

where $\vec{b}$ is orthogonal to the unperturbed light trajectory and is directed toward the point-like lens. Inserting this result in Eq. (2.9) we find[1]

$$\vec{\alpha} = \frac{2}{c^2}\int_{-\infty}^{\infty} \nabla_\perp \Phi \mathrm{d}z = \frac{4GM}{c^2 b}\frac{\vec{b}}{|\vec{b}|} = \frac{2R_s}{b}\frac{\vec{b}}{|\vec{b}|}, \qquad (2.12)$$

where $R_{\mathrm{s}} = \frac{2GM}{c^2}$ is the Schwarzschild radius for a body of mass M. Thus the absolute value of the deflection angle is $\alpha = 2R_{\mathrm{S}}/b$. For the Sun the Schwarzschild radius is

---

[1] Since x_{S} and x_{O} are much bigger as compared to the region where the lens perturbs the trajectory of light, one can approximate x_{S} and x_{O} by $\pm\infty$.

2.95 km, whereas its physical radius is $6.96 \cdot 10^5$ km. A light ray which just grazes the solar surface is deflected by an angle of $1.75''$.

2.1.3 Thin Lens Approximation

From the above considerations one sees that the main contribution to the light deflection comes from the region $\Delta z \simeq \pm b$ around the lens. This distance is typically much smaller than the distance between observer and lens D_d and the distance between lens and source D_{ds}. The lens can thus be assumed to be thin compared to the full length of the light trajectory.

We consider the mass of the lens, for instance a galaxy cluster, projected on a plane perpendicular to the line of sight (between observer and lens) and going through the center of the lens. This plane is usually referred to as the *lens plane*, and similarly one can define the *source plane*. The projection of the lens mass on the lens plane is obtained by integrating the mass density ρ along the direction perpendicular to the lens plane

$$\Sigma(\vec{\xi}) = \int \rho(\vec{\xi}, z) \mathrm{d}z, \tag{2.13}$$

where $\vec{\xi}$ is a two-dimensional vector in the lens plane and z is the distance from the plane.

The deflection angle at the point $\vec{\xi}$ is then given by summing over the deflection due to all mass elements in the plane as follows

$$\vec{\alpha}(\vec{\xi}) = \frac{4G}{c^2} \int \frac{(\vec{\xi} - \vec{\xi}')\Sigma(\vec{\xi}')}{|\vec{\xi} - \vec{\xi}'|^2} \mathrm{d}^2\vec{\xi}'. \tag{2.14}$$

Note that for one mass element (1 particle), $\Sigma(\vec{\xi}') = M\delta^2(\vec{\xi}' - \vec{\xi}_0)$.

In the general case the deflection angle is described by a two-dimensional vector, however in the special case that the lens has circular symmetry one can reduce the problem to a one-dimensional situation. Then the deflection angle $\vec{\alpha}$ is a vector directed toward the center of the symmetry with absolute value given by

$$\alpha = \frac{4GM(\xi)}{c^2\xi}, \tag{2.15}$$

where ξ is the distance from the center of the lens and $M(\xi)$ is the total mass inside a radius ξ from the center, defined as

$$M(\xi) = 2\pi \int_0^{\xi} \Sigma(\xi')\xi' \mathrm{d}\xi'. \tag{2.16}$$

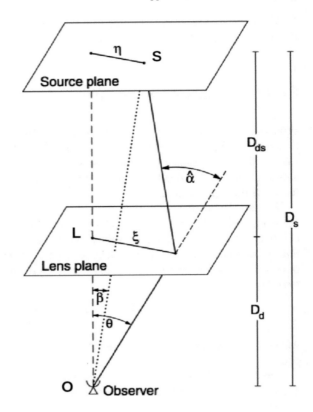

Fig. 2.1 Geometry of a gravitational lens

2.1.4 *Lens Equation*

The geometry for a typical gravitational lens is given in Fig. 2.1. A light ray coming from a source S, laying in the source plane, is deflected by the lens L by an angle α, with ξ as the impact parameter, and reaches the observer O.

We arbitrarily define β to be the angle between the optical axis (line going through the observer and the lens positions) and the true source position, whereas θ is the angle between the optical axis and the image position. As mentioned before, the distances along the line of sight between observer and lens, lens and source and observer and source are D_d, D_{ds} and D_s, respectively.

Assuming small angles, we can easily derive $\vec{\theta} D_s = \vec{\beta} D_s + \vec{\alpha} D_{ds}$ from the geometry of the problem. Rearranging terms we get

$$\vec{\beta} = \vec{\theta} - \vec{\alpha}(\vec{\theta}) \frac{D_{ds}}{D_s},\tag{2.17}$$

which is known as the *lens equation*. This is a nonlinear equation, and it is possible to have several images $\vec{\theta}$ corresponding to a single source position $\vec{\beta}$.

The lens equation (2.17) can also be derived using Fermat's principle, which is identical to the classical one in geometrical optics but with the refraction index as defined in Eq. (2.1). The light trajectory is given then by the variational principle

$$\delta \int n(\vec{x}) \mathrm{d}\vec{x} = 0 \qquad (2.18)$$

as we did in Eq. (2.5). It expresses the fact that the light trajectory will be such that the traveling time will be extremal.

Let us consider a light ray emitted from the source S at time $t = 0$. The light ray will proceed straight till it reaches the lens, located at L in the lens plane, where it will be deflected and then proceed again straight to the observer O. We can thus compute

$$t = \int_{x_S}^{x_O} \frac{n}{c} \mathrm{d}x = \frac{1}{c} \int_{x_S}^{x_O} \left(1 - \frac{2\Phi}{c^2}\right) \mathrm{d}x = \frac{l}{c} - \frac{2}{c^3} \int_{x_S}^{x_O} \Phi \mathrm{d}x, \qquad (2.19)$$

where l is the Euclidean distance SLO. From the geometry of the problem we obtain

$$l = \sqrt{(\vec{\xi} - \vec{\eta})^2 + D_{ds}^2} + \sqrt{\vec{\xi}^2 + D_d^2} \simeq D_{ds} + D_d + \frac{1}{2D_{ds}}(\vec{\xi} - \vec{\eta})^2 + \frac{1}{2D_d}\vec{\xi}^2,$$
$$(2.20)$$

where $\vec{\eta}$ is a two dimensional vector on the source plane, and where we did a simple Taylor expansion.

The second term of Eq. (2.19) containing Φ has to be integrated along the light trajectory. If we assume a point-like lens then the Newtonian potential takes the form $\Phi = -GM/|\vec{x}|$, we get

$$\int_{x_S}^{x_L} \frac{2\Phi}{c^3} \mathrm{d}x = \frac{2GM}{c^3}\left(\ln \frac{|\vec{\xi}|}{2D_{ds}} + \frac{\vec{\xi}(\vec{\eta} - \vec{\xi})}{|\vec{\xi}|D_{ds}} + \mathcal{O}\left(\frac{(\vec{\eta} - \vec{\xi})^2}{D_{ds}}\right)\right), \qquad (2.21)$$

and a similar expression for $\int_{x_L}^{x_O} 2\Phi/c^3 \mathrm{d}x$. While for the thin lens approximation with a surface density $\Sigma(\xi)$ given by Eq. (2.13) we obtain, neglecting higher order contributions,

$$\int_{x_S}^{x_O} \frac{2\Phi}{c^3} \mathrm{d}x = \frac{4G}{c^3} \int \mathrm{d}^2\vec{\xi}' \Sigma(\vec{\xi}') \ln \left(\frac{|\vec{\xi} - \vec{\xi}'|}{\xi_0}\right), \qquad (2.22)$$

where ξ_0 is a characteristic length in the lens plane and the right-hand side term is defined up to a constant.

The difference in arrival time between a photon deflected by the lens and one not deflected at all is obtained by subtracting the travel time without deflection from S to O to the deflected travel time given by Eq. (2.19). Thus the two first terms of the expansion in Eq. (2.20) disappear. This way one gets

$$c\Delta t = \hat{\phi}(\vec{\xi}, \vec{\eta}) + \text{const.}, \tag{2.23}$$

where $\hat{\phi}(\vec{\xi}, \vec{\eta})$ is the *Fermat potential* and it is

$$\hat{\phi}(\vec{\xi}, \vec{\eta}) = \frac{D_{\mathrm{d}} D_{\mathrm{s}}}{2 D_{\mathrm{ds}}} \left(\frac{\vec{\xi}}{D_{\mathrm{d}}} - \frac{\vec{\eta}}{D_{\mathrm{s}}} \right)^2 - \hat{\psi}(\vec{\xi}), \tag{2.24}$$

where

$$\hat{\psi}(\vec{\xi}) = \frac{4G}{c^2} \int \mathrm{d}^2\vec{\xi}' \Sigma(\vec{\xi}') \ln \left(\frac{|\vec{\xi} - \vec{\xi}'|}{\xi_0} \right), \tag{2.25}$$

is the *deflection potential* and it doesn't depend on $\vec{\eta}$.

The Fermat's principle can thus be written as $\mathrm{d}\Delta t / \mathrm{d}\vec{\xi} = 0$, inserting Eq. (2.23) we get again the lens equation

$$\vec{\eta} = \frac{D_{\mathrm{s}}}{D_{\mathrm{d}}} \vec{\xi} - D_{\mathrm{ds}} \vec{\alpha}(\vec{\xi}), \tag{2.26}$$

where $\vec{\alpha}$ is defined in Eq. (2.14). Defining $\vec{\beta} = \vec{\eta}/D_{\mathrm{s}}$ and $\vec{\theta} = \vec{\xi}/D_{\mathrm{d}}$ we recover Eq. (2.17). We can also rewrite Eq. (2.26) as

$$\nabla_{\vec{\xi}} \hat{\phi}(\vec{\xi}, \vec{\eta}) = 0, \tag{2.27}$$

which is an equivalent form of the Fermat principle. The arrival time delay of light rays coming from two different images located at $\vec{\xi}^{(1)}$ and $\vec{\xi}^{(2)}$ from the same source $\vec{\eta}$ is given by

$$c(t_1 - t_2) = \hat{\phi}(\vec{\xi}^{(1)}, \vec{\eta}) - \hat{\phi}(\vec{\xi}^{(2)}, \vec{\eta}). \tag{2.28}$$

It can be shown that the *magnification* factor μ for a given image θ is given by (see, e.g., [5])

$$\mu = \frac{1}{\det A(\theta)}, \tag{2.29}$$

where $A(\theta) = \frac{\mathrm{d}\vec{\beta}}{\mathrm{d}\theta}$ and $A_{ij} = \frac{\mathrm{d}\beta_i}{\mathrm{d}\theta_j}$.

2.1.5 Schwarzschild Lens

For a lens with axial symmetry, using the deflection angle obtained in Eq. (2.15), we get the following lens equation

$$\beta(\theta) = \theta - \frac{D_{\mathrm{ds}}}{D_{\mathrm{s}} D_{\mathrm{d}}} \frac{4 G M(\theta)}{c^2 \theta}. \tag{2.30}$$

From this equation we see that the image of a perfectly aligned source ($\beta = 0$) is a ring if the lens is supercritical.[2] By setting $\beta = 0$ in Eq. (2.30) we get the radius of the ring

$$\theta_E = \left(\frac{4GM(\theta_E)}{c^2} \frac{D_{ds}}{D_d D_s} \right)^{\frac{1}{2}}, \tag{2.31}$$

which is called *Einstein radius*.[3] The Einstein radius depends not only on the characteristics of the lens but also on the various distances.

The Einstein radius sets a natural scale for the angles entering the description of the lens. Indeed, for multiple images the typical angular separation between the different images turns out to be of order $2\theta_E$. Moreover, sources with angular distance smaller than θ_E from the optical axis of the system get magnified quite substantially, whereas sources which are at a distance much greater than θ_E are only weakly magnified.

A particular case of a lens with axial symmetry is the Schwarzschild lens, for which $\Sigma(\vec{\xi}) = M\delta^{(2)}(\vec{\xi})$ and thus $m(\theta) = \theta_E^2$. If the source is also point-like, this way we get the following lens equation

$$\beta = \theta - \frac{\theta_E^2}{\theta}. \tag{2.32}$$

This equation has two solutions

$$\theta_\pm = \frac{1}{2} \left(\beta \pm \sqrt{\beta^2 + 4\theta_E^2} \right). \tag{2.33}$$

Therefore, there will be two images of the source, one located inside the Einstein radius and the other outside. For a lens with axial symmetry the magnification is given by

$$\mu = \frac{\theta}{\beta} \frac{d\theta}{d\beta}. \tag{2.34}$$

For the Schwarzschild lens, which is a limiting case of an axial symmetric one, we can use β given by Eq. (2.32) and obtain the magnification of the two images

$$\mu_\pm = \left(1 - \left(\frac{\theta_E}{\theta_\pm} \right)^4 \right)^{-1} = \frac{u^2 + 2}{2u\sqrt{u^2 + 4}} \pm \frac{1}{2}, \tag{2.35}$$

where $u = r/R_E$ is the ratio between the impact parameter $r = \beta D_d$ and the Einstein radius given by $R_E = \theta_E D_d$. The parameter u can thus also be expressed as $u = \beta/\theta_E$. Since $\theta_- < \theta_E$ we have that $\mu_- < 0$. The negative sign for the magnifications indicates that the parity of the image is inverted with respect to the source.

---

[2] A lens which has $\Sigma > \Sigma_{crit} = \frac{c^2}{4\pi G} \frac{D_s}{D_d D_{ds}}$ somewhere in it is defined as supercritical, and has in general multiple images.

[3] Instead of the angle θ_E one often uses $R_E = \theta_E D_d$.

The total amplification is given by the sum of the absolute values of the magnifications for each image

$$\mu = |\mu_+| + |\mu_-| = \frac{u^2 + 2}{u\sqrt{u^2 + 4}}. \tag{2.36}$$

If $r = R_E$ then we get $u = 1$ and $\mu = 1.34$, which corresponds to an increase of the apparent magnitude of the source of $\Delta m = -2.5 \log \mu = -0.32$. For lenses with a mass of the order of a solar mass and which are located in the halo of our galaxy the angular separation between the two images is far too small to be observable. Instead, one observes a time-dependent change in the brightness of the source star as the lensing object is moving on its orbit. This situation is also referred to as *microlensing*.

For the many applications of gravitational lensing in cosmology and astrophysics we refer to the now vast available literature, e.g., [7].

2.2 General Relativistic Stellar Structure Equations

2.2.1 *Introduction*

Although the applications of general relativity to astronomical problems started already with the works of Einstein himself, among which the deflection of light and the precession of Mercury's perihelion, it was not until the discovery of pulsars[4] in 1967 that a great deal of interest was directed toward the impact of the theory on stellar structure. The first pioneering theoretical works were done already 30 years earlier by Landau [8] and later on by Oppenheimer and Volkoff [9]. Once the existence of neutron stars and white dwarfs was definitively established a modeling work for these objects started, which lead to our contemporary view of these objects.

Before proceeding, it is useful to very briefly remember how a star can evolve into either a *white dwarf, neutrons star* or *black hole* according to our present days best models. A star with mass $M < 4\,M_\odot$ cannot start the carbon fusion process and thus evolves into a white dwarf by a collapse, where the relatively gentle mass ejection forms a planetary nebulae. The white dwarf is the stellar core left behind that is supported by electron degeneracy pressure, due to Pauli exclusion principle.

Throughout the life of heavier stars, a core of iron is formed. At the end of the star's life, the outer layers collapse. The gravitational energy of the layers make it possible to fuse or fission the iron core. The collapse hence continues past the electron degeneracy pressure until the neutron degeneracy pressure stops it. The layers rebounce on the core of the star, now made mostly of neutrons, and form a supernovae. For stars with masses 10–$30 M_\odot$, a neutron star with mass $1.4 M_\odot$ to

---

[4] A pulsar is a highly magnetized, rotating neutron star or white dwarf.

2–$3M_\odot$ is remnant. A neutron star is hence almost entirely composed of neutrons and held by neutron degeneracy pressure. Whereas for white dwarfs the maximum possible mass is well established, this is not yet the case for a neutron star, although the range for the maximum mass is getting more and more tight. For sufficiently massive stars, the core will accrete too much mass to be stable and will finally collapse into a black hole.

2.2.2   *General Relativistic Stellar Structure Equations*

We derive the general relativistic stellar structure equations for non-rotating, static and spherically symmetric compact stars (see, e.g., [10]).

For such stars the metric is the Schwarzschild metric (1.25) or (1.41) and has the following form when defining $B(r) \equiv e^{2a(r)}$, $A(r) \equiv e^{2b(r)}$

$$ds^2 = -e^{2a(r)}dt^2 + e^{2b(r)}dr^2 + r^2(d\theta^2 + \sin^2\theta d\phi^2), \tag{2.37}$$

where we use units in which $c = 1$. Moreover, we adopt the signature $(-, +, +, +)$, whereas in (1.25) or (1.41) we had the convention with the signature $(+, -, -, -)$.

The Einstein tensor corresponding to the above metric is given by (as can easily be derived from Eqs. 1.32–1.35)

$$G^0{}_0 = -\frac{1}{r^2} + e^{-2b}\left(\frac{1}{r^2} - \frac{2b'}{r}\right), \tag{2.38}$$

$$G^1{}_1 = -\frac{1}{r^2} + e^{-2b}\left(\frac{1}{r^2} + \frac{2a'}{r}\right), \tag{2.39}$$

$$G^2{}_2 = G^3{}_3 = e^{-2b}\left(a'^2 - a'b' + a'' + \frac{a' - b'}{r}\right), \tag{2.40}$$

$$G^\mu{}_\nu = 0 \text{ for all the other components.} \tag{2.41}$$

Here the $'$ indicate the derivative with respect to r.

We choose to model the matter and energy in the star as a perfect fluid neglecting any anisotropic stresses and heat conduction. The energy-momentum tensor of an isotropic perfect fluid is

$$T_{\mu\nu} = \left(\rho + \frac{p}{c^2}\right)u_\mu u_\nu - p g_{\mu\nu}, \tag{2.42}$$

where u^μ is the four-velocity of the fluid and ρ and p are the energy density and pressure as measured in the rest frame. The field equations $G^\mu{}_\nu = 8\pi G T^\mu{}_\nu$ (with $c = 1$) are

$$\frac{1}{r^2} - e^{-2b}\left(\frac{1}{r^2} - \frac{2b'}{r}\right) = 8\pi G\rho, \tag{2.43}$$

$$\frac{1}{r^2} - e^{-2b}\left(\frac{1}{r^2} + \frac{2a'}{r}\right) = -8\pi Gp. \tag{2.44}$$

Defining $u/r \equiv e^{-2b(r)}$, the first field equation can be rewritten as

$$u' = -8\pi G\rho r^2 + 1, \tag{2.45}$$

and through integration we obtain

$$u = r - 2GM(r), \tag{2.46}$$

where

$$M(r) = 4\pi \int_0^r \rho(\tilde{r})\tilde{r}^2 d\tilde{r}, \tag{2.47}$$

and thus

$$e^{-2b} = 1 - \frac{2GM(r)}{r}. \tag{2.48}$$

Now if we subtract the second field equation from the first one, we obtain

$$e^{-2b}(a' + b') = 4\pi G(\rho + p)r, \tag{2.49}$$

and after integration

$$a = -b + 4\pi G \int_\infty^r e^{2b(\tilde{r})}\tilde{r}(\rho + p)(\tilde{r})d\tilde{r}. \tag{2.50}$$

If ρ and p are known one can then determine the gravitational field.

An additional useful relation can be derived from the conservation law

$$T^{\mu\nu}{}_{;\nu} = 0 \Leftrightarrow T^{\mu\nu}{}_{,\nu} + \Gamma^\mu_{\lambda\nu}T^{\lambda\nu} + \Gamma^\nu_{\lambda\nu}T^{\mu\lambda} = 0 \tag{2.51}$$

$$\Leftrightarrow \frac{1}{\sqrt{|g|}}\frac{\partial}{\partial x^\nu}(\sqrt{|g|}T^{\mu\nu}) + \Gamma^\mu_{\mu\lambda}T^{\lambda\nu} = 0. \tag{2.52}$$

Inserting $T_{\mu\nu}$ for $\mu = 1$ in the above expression and using the fact that $g^{\mu\nu}{}_{;\nu} = 0$, we obtain

$$a' = \frac{-p'}{(p + \rho)}. \tag{2.53}$$

On the other hand, combining Eqs. (2.47), (2.48) and (2.50) we recover

$$a' = \frac{G}{1 - \frac{2GM(r)}{r}} \left(\frac{M(r)}{r^2} + 4\pi r p \right). \tag{2.54}$$

Comparing this equation with Eq. (2.53) we get the *Tolman–Oppenheimer–Volkoff* (TOV) equation

$$\frac{dp}{dr} = -\frac{GM(r)\rho}{r^2} \left(1 + \frac{p}{\rho c^2} \right) \left(1 + \frac{4\pi r^3 p}{Mc^2} \right) \left(1 - \frac{2GM(r)}{c^2 r} \right)^{-1}, \tag{2.55}$$

where the pressure vanishes at the stellar radius R and the gravitational mass is given by

$$M(r) = 4\pi \int_0^r \rho(\tilde{r})\tilde{r}^2 d\tilde{r}. \tag{2.56}$$

The TOV equation is the generalization of the *hydrostatic equilibrium* equation

$$\frac{dp}{dr} = -\frac{GM(r)\rho(r)}{r^2}, \tag{2.57}$$

of the non-relativistic Newtonian theory. Within general relativity the radial pressure gradient increases for the following three reasons:

- Since gravity also acts on the pressure p, the density ρ is replaced by $(\rho + \frac{p}{c^2})$ (first term in brackets of TOV equation).
- Since the pressure also acts as a source of the gravitational field, there is a term proportional to p in addition to $M(r)$ (second term in brackets of TOV equation).
- The gravitational force increases faster than $1/r^2$ of the Newtonian case, and indeed in general relativity this quantity is replaced by $r^{-2}(1 - 2GM(r)/c^2 r)^{-1}$ (third term in brackets of TOV equation).

In order to construct a stellar model, one needs an *equation of state* $p = p(\rho)$. If this is known[5] the density profile of the star is determined by its central energy density $\rho_c \equiv \rho(r = 0)$.

The general relativistic stellar structure equations are (putting $c = 1$)

$$-p' = \frac{G(\rho + p)(M(r) + 4\pi r^3 p)}{r^2(1 - 2GM(r)/r)}, \tag{2.58}$$

---

[5] One usually makes some assumptions and chooses the equation of state accordingly. In general the equation of state is of the form $p = p(\rho, T, \{X_i\})$, where $\{X_i\}$ is the set of the abundances of all elements i and T their temperature. These dependencies increase the complexity of the model. Furthermore one has to consider the temperature structure of the star, its abundance ratios governed by nuclear physics and the various heat transport mechanisms.

$$M' = 4\pi r^2 \rho, \tag{2.59}$$

$$a' = \frac{G}{1 - 2GM(r)/r}\left(\frac{M(r)}{r^2} + 4\pi rp\right). \tag{2.60}$$

The initial condition $M(r = 0) = 0$ is obtained from Eq. (2.56), the other two boundary conditions are $p(r = 0) = p(\rho_c)$ and $e^{2a(R)} = 1 - 2GM(R)/R$ assuming a Schwarzschild metric outside the star.

2.2.3   Interpretation of the Gravitational Mass M

We want to compare the gravitational mass M of Eq. (2.56) with the baryonic mass $M_0 = Nm_N$, where N is the total number of nucleons in the star and m_N the mass of the nucleon. To get this quantity we have to integrate the baryon number density $n(r)$ over the volume as follows

$$N = 4\pi \int_0^R n(r)e^{b(r)}r^2 dr = 4\pi \int_0^R \frac{n(r)r^2 dr}{\sqrt{1 - \frac{2GM(r)}{r}}}. \tag{2.61}$$

Note that here the volume element is not Minkowskian (which would just be $dV = dxdydz$) but, according to the metric defined in Eq. (2.37), given by $dV = e^{b(r)}r^2 dr \sin\theta d\theta d\phi$.[6]

The proper internal material energy density is defined as

$$\varepsilon(r) \equiv \rho(r) - m_N n(r), \tag{2.62}$$

and the total internal energy is correspondingly

$$E = M - M_0 = M - m_N N. \tag{2.63}$$

We use Eqs. (2.56) and (2.61) to decompose $E = T + V$ as follows

$$E = 4\pi \int_0^R r^2\left(\rho(r) - \frac{n(r)m_N}{\sqrt{1 - \frac{2GM(r)}{r}}}\right)dr \tag{2.64}$$

$$= 4\pi \int_0^R \frac{r^2\varepsilon(r)}{\sqrt{1 - \frac{2GM(r)}{r}}}dr + 4\pi \int_0^R r^2\rho(r)\left(1 - \frac{1}{\sqrt{1 - \frac{2GM(r)}{r}}}\right)dr \equiv T + V. \tag{2.65}$$

---

[6] Here $e^{b(r)}r^2 dr$ denotes the radial part, $\sin\theta d\theta d\phi = d\Omega$ the angular one.

In order to find the connection with Newtonian theory, we expand the square roots in T and V, assuming that $GM(r)/r \ll 1$. This leads us to

$$T = 4\pi \int_0^R r^2 \varepsilon(r) \left(1 + \frac{GM(r)}{r} + \cdots \right) dr, \tag{2.66}$$

$$V = -4\pi \int_0^R r^2 \rho(r) \left(\frac{GM(r)}{r} + \frac{3G^2 M^2(r)}{2r^2} + \cdots \right) dr. \tag{2.67}$$

The leading terms in T and V are the Newtonian values for the internal and gravitational energy of the star. Note that the extremely small value G^2 appears in the second term of the Taylor expansion for the potential.

2.2.4 The Interior of Neutron Stars

The physics of the interior of a neutron star is extremely complicated. A major problem is to establish a reliable equation of state. We assume that the neutron star is made up by an *ideal* mixture of nucleons and electrons in β-equilibrium.

In a system where quantum degeneracy is prevalent, the states in momentum space occupy a Fermi sphere of radius p_F. The energy density ρ, number density n and pressure p of an ideal Fermi gas at temperature $T = 0$ are given (in units of $c = 1$) by:

$$\rho = \frac{8\pi}{(2\pi\hbar)^3} \int_0^{p_F} \sqrt{p^2 + m_N^2}\, p^2 dp, \tag{2.68}$$

$$n = \frac{8\pi}{(2\pi\hbar)^3} \int_0^{p_F} p^2 dp, \tag{2.69}$$

$$p = \frac{1}{3} \frac{8\pi}{(2\pi\hbar)^3} \int_0^{p_F} \frac{p^2}{\sqrt{(p^2 + m_N^2)}} p^2 dp, \tag{2.70}$$

where

$$\rho_0 = \frac{8\pi m_N^4 c^3}{3(2\pi\hbar)^3} = 6.11 \cdot 10^{15} \frac{\text{g}}{\text{cm}^3}. \tag{2.71}$$

In 1939, Oppenheimer and Volkoff [9] performed the first neutron star calculations assuming that such objects are entirely made of a gas of non-interacting relativistic neutrons, thus using the equation of state as given in previous Eqs. (2.68)–(2.70)

$$\rho = \rho\left(\frac{p_F}{m_N}\right), \quad p = p\left(\frac{p_F}{m_N}\right) \Rightarrow p = p(\rho). \tag{2.72}$$

This equation of state is extremely "soft" meaning that very little additional pressure is gained with increasing density, and it predicts a maximum neutron star mass of just $M_{max} = 0.71\, M_\odot$ with radius $R_{max} = 9.3$ km. Since the neutron star's mass is a function of the central density $M = M(\rho_c)$, one finds $\rho_{c,max} = 4 \cdot 10^{15}$ g/cm^3. For values of $\rho_c > \rho_{c,max}$ the star is unstable.

For more realistic equations of state, when other particles and interactions among the neutrons are considered, the star's maximum mass increases from $0.71\, M_\odot$ to 2–$3\, M_\odot$. Indeed most observed neutron stars have masses around $\simeq 1.4\, M_\odot$ [11].

We can now use the listed properties of a neutron star to obtain an estimate of the gravitational redshift of a photon emitted at the surface of the neutron star. We find (with $\vec{r}_A$ being the source position, in this case the surface of the neutron star, and $\vec{r}_B$ observer's position):

$$z = \frac{\Delta\lambda}{\lambda} = \sqrt{\frac{g_{00}(\vec{r}_B)}{g_{00}(\vec{r}_A)}} - 1 = \left(1 - \frac{2GM_{max}}{R_{max}}\right)^{-\frac{1}{2}} - 1 \simeq 0.13. \tag{2.73}$$

Furthermore, a rapidly rotating neutron star can also emit gravitational waves. Its rapid rotational motion flattens its structure and generates a quadrupole radiation (see Chap. 3). The flattening of the neutron star is given by the *ellipticity*

$$\epsilon \equiv \frac{r_{equatorial} - r_{polar}}{r_{average}}, \tag{2.74}$$

where $r_{average} = \frac{r_{equatorial} + r_{polar}}{2}$. One can use the flattening ϵ together with the moment of inertia I of the neutron star to calculate the emitted energy per unit time (see derivation in Sect. 3.1)

$$P = \dot{E}_{grav} = -\frac{32}{5}\frac{G}{c^5}I^2\epsilon^2\Omega^6. \tag{2.75}$$

Using LIGO and Virgo data [12] (for more recent limits see [13]) it can be excluded that neutron stars have an ellipticity $\epsilon > 10^{-5}\sqrt{10^{38}\,\text{kg}\,\text{m}^2/I}$ within 100 pc of Earth for frequencies above 55 Hz.

2.3 Time Delay of Radar Echoes

In 1964 Shapiro [14] proposed a new test of GR consisting of a measurement of the time delay of radar signals transmitted from Earth through a region near the Sun to another planet or spacecraft and reflected back to Earth. Since the radar signal is affected by the gravitational field of the Sun, it will return to Earth with a time delay.

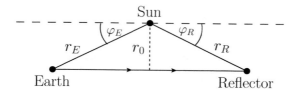

Fig. 2.2 Sketch of the system as described in the text. The angles φ_E and φ_R and the distances r_E, r_R describe the position of the Earth and the reflector relative to the Sun. The whole setup is chosen to lie in the $\theta = \frac{\pi}{2}$-plane

The radar signal is emitted from Earth and then sent back by a reflector as sketched in Fig. 2.2. We assume that the light trajectories and the Earth and reflector to lie in the plane corresponding to $\theta = \frac{\pi}{2}$. The trajectory of the signal is curved. This is not shown in the figure because the curvature is, of course, only a tiny effect and the light rays are thus almost straight lines. We compute the time that the signal needs to get from Earth to the reflector. To do this we use Eq. (1.57) for the orbit in a static isotropic gravitational field with $\varepsilon = 0$ (since $m = 0$):

$$A \left(\frac{dr}{d\lambda} \right)^2 + \frac{l^2}{r^2} - \frac{F^2}{B} = 0. \tag{2.76}$$

If we divide this expression by l^2 and use $x^0 = ct$, we find (with Eq. 1.54)

$$\frac{dr}{d\lambda} = \frac{1}{c} \frac{dr}{dt} \frac{dx^0}{d\lambda} = \frac{1}{c} \frac{dr}{dt} \frac{F}{B}. \tag{2.77}$$

This yields

$$\frac{AF^2}{c^2 B^2 l^2} \left(\frac{dr}{dt} \right)^2 + \frac{1}{r^2} - \frac{F^2}{Bl^2} = 0. \tag{2.78}$$

For the minimal distance r_0 (from the Sun) it holds that

$$\frac{dr}{dt} = 0 \quad \Rightarrow \quad \frac{F^2}{l^2} = \frac{B(r_0)}{r_0^2}. \tag{2.79}$$

We insert this into Eq. (2.78) to obtain

$$\frac{A}{c^2 B} \left(\frac{dr}{dt} \right)^2 + \frac{r_0^2}{r^2} \frac{B(r)}{B(r_0)} - 1 = 0. \tag{2.80}$$

This differential equation is solved by the following integral:

$$t(r, r_0) = \frac{1}{c} \int_{r_0}^{r} dr' \sqrt{\frac{A}{B}} \left[1 - \frac{r_0^2}{r'^2} \frac{B(r')}{B(r_0)} \right]^{-1/2}, \tag{2.81}$$

where $t(r, r_0)$ is the time that the radar signal needs to travel from r_0 to r. Note that this is the time which would be shown by a clock at rest at infinity (as space is asymptotically Minkowskian at infinity). This actually forces us to introduce a correction since our clock rests at Earth, not at infinity. However, the correction which is needed to compensate this effect is much smaller than the time delay and can thus be neglected.

Using the Robertson expansion from Eq. (1.36),

$$A(r) = 1 + \gamma \frac{2a}{r} + \cdots, \qquad B(r) = 1 - \frac{2a}{r} + \cdots, \tag{2.82}$$

we get

$$1 - \frac{r_0^2}{r^2} \frac{B(r)}{B(r_0)} = 1 - \frac{r_0^2}{r^2} \left[1 + 2a \left(\frac{1}{r_0} - \frac{1}{r} \right) \right]$$
$$= \left[1 - \frac{r_0^2}{r^2} \right] \left[1 - \frac{2ar_0}{r(r + r_0)} \right]. \tag{2.83}$$

Inserting Eqs. (2.82) and (2.83) into (2.81) and expanding, we get

$$t(r, r_0) \simeq \frac{1}{c} \int_{r_0}^{r} dr' \left[1 - \frac{r_0^2}{r'^2} \right]^{-1/2} \left(1 + \frac{ar_0}{r'(r' + r_0)} + (1 + \gamma) \frac{a}{r'} \right)$$
$$= \frac{\sqrt{r^2 - r_0^2}}{c} + \frac{a}{c} \sqrt{\frac{r - r_0}{r + r_0}} + (1 + \gamma) \frac{a}{c} \log \left(\frac{r + \sqrt{r^2 - r_0^2}}{r_0} \right). \tag{2.84}$$

The first term $\frac{\sqrt{r^2 - r_0^2}}{c}$ corresponds to the traveling time assuming a straight trajectory in Euclidean space as can easily be seen from Fig. 2.2 and Pythagoras' theorem. The other terms account for the general relativistic time delay due to the gravitational field of the Sun. For the system drawn in Fig. 2.2, the total delay is

$$\delta t = 2 \left[t(r_E, r_0) + t(r_R, r_0) - \frac{\sqrt{r_E^2 - r_0^2}}{c} - \frac{\sqrt{r_R^2 - r_0^2}}{c} \right], \tag{2.85}$$

where the factor of 2 accounts for the fact that the signal travels from Earth to the reflector and back again.

Significant delays occur if the radar signal passes nearby the Sun, i.e., if r_0 is of the order of some Sun radii. In this case we have $r_E, r_R \gg R_\odot$ and also $r_E, r_R \gg r_0$. With these approximations in Eq. (2.85) [14] we get:

$$\delta t \simeq \frac{4a}{c}\left[1 + \left(\frac{1+\gamma}{2}\right)\log\frac{4r_E r_R}{r_0^2}\right]. \tag{2.86}$$

We see that δt is maximal if the signal just grazes the surface of the Sun, i.e., $r_0 = R_\odot$.

To get the orders of magnitude, we use $r_E \sim r_R \sim 10^8$ km and $R_\odot \simeq 7 \cdot 10^5$ km. This yields $\frac{2a}{c} = \frac{2GM_\odot}{c^3} \simeq 10^{-5}$ s and thus

$$\delta t_{max} = \frac{4a}{c}\left[1 + \left(\frac{1+\gamma}{2}\right)\log\frac{4r_E r_R}{R_\odot^2}\right]$$
$$\simeq 2 \cdot 10^{-4} \text{ s.} \tag{2.87}$$

Performing the measurements has been a very difficult task since the distances r_E and r_R were not known with sufficient precision. Nevertheless, in the seventies these measurements were performed using Venus and Mercury and later again using spacecraft as reflectors (e.g., the Vikings which landed on Mars or later the Cassini spacecraft [15]). The measurements lead to the following results

$$\text{Vikings:}\quad \gamma = 1.000 \pm 0.001,$$
$$\text{Cassini:}\quad \gamma = 1 + (2.1 \pm 2.3) \cdot 10^{-5}. \tag{2.88}$$

Another experimental verification of the Shapiro delay is, e.g., given by the measurement of the PSR J1614-2230 system. It consists of a pulsar which emits signals in very regular time intervals, and a white dwarf that orbits the pulsar. When the white dwarf is in front of the pulsar and the light of the pulsar arrives at Earth by passing close to the white dwarf, the signal arrives with a delay. Measuring the Shapiro delay [16], one can infer the mass of the white dwarf to be $0.500 \pm 0.006 M_\odot$. With an orbital period of 8.7 days, this yields a neutron star mass of $1.97 \pm 0.04 M_\odot$. This result is important for the modeling of neutron stars as it indicates that the maximum neutron star mass is higher then $1.4 M_\odot$, which is the usually observed value.

2.4   Precession

Consider a particle with a "classical" angular momentum (for instance the intrinsic angular momentum of a rigid body like a gyroscope). In the local inertial frame system (IS) in which the body is at rest, the spin (i.e., angular momentum) is given by $\vec{S} = S^i \vec{e}_i$. To the three-vector S^i we assign a Lorentz vector S^α. Consider now a locally inertial coordinate frame IS' which is momentarily at rest with respect to the rigid body (or particle):

$$S'^\alpha = (0, S'^i). \tag{2.89}$$

We can transform this to an arbitrary inertial frame system IS by means of a Lorentz transformation. In the rest frame IS', the velocity of the body is

$$u'^\alpha = (c, \vec{0}).$$

Therefore we have in IS'

$$u'_\alpha S'^\alpha = 0.$$

Since this quantity is a Lorentz scalar, $u_\alpha S^\alpha = 0$ in any arbitrary inertial frame IS.

Consider first the case without any forces acting on the particle and no torque acting on its spin. In an arbitrary IS we have

$$\frac{dS^\alpha}{d\tau} = 0. \tag{2.90}$$

We define the Riemann vector

$$S^\mu \equiv \frac{\partial x^\mu}{\partial \xi^\alpha} S^\alpha, \tag{2.91}$$

which describes the transition from the coordinate system IS with coordinates (ξ^α) to a general coordinate system (x^μ). According to the covariance principle, the generalization of Eq. (2.90) reads (with $\frac{D}{d\tau}$ denoting the covariant derivative)

$$\frac{DS^\mu}{d\tau} = 0 \qquad \Leftrightarrow \qquad \frac{dS^\mu}{d\tau} = -\Gamma^\mu{}_{\nu\lambda} u^\nu S^\lambda. \tag{2.92}$$

This equation describes the spin precession in a gravitational field. The condition $u_\alpha S^\alpha = 0$ becomes $u_\mu S^\mu = 0$ in the general coordinate system. Note that $S_\mu S^\mu = $ const., which implies that (2.92) describes the *rotation* or *precession* of the spin vector. Since we assumed no external forces, Eq. (2.92) contains only gravitational effects. Therefore, Eq. (2.92) describes the precession of the spin of a particle which is freely falling in a gravitational field as, for example, the precession of a rotating satellite (or gyroscope) in Earth's gravitational field.

If there are other external forces f^μ besides gravity, then one finds instead of Eq. (2.92)

$$\frac{DS^{\nu}}{d\tau} = -\frac{1}{c^2}\frac{Du^{\mu}}{d\tau}S_{\mu}u^{\nu}, \tag{2.93}$$

which is also called *Fermi transport*. It describes the spin precession of an accelerated particle on which a gravitational field acts (in general relativity Newton's equation reads: $Du^{\mu}/d\tau = f^{\mu}/m$. The special case $f^{\mu} = 0$ corresponds to parallel transport). Based on the above considerations, we shall study the following effects (in the gravitational field of the Earth):

1. **Geodetic precession**: the precession of a freely falling gyroscope. In order to simplify the analysis we will assume the gravitational field to be isotropic and static.
2. **Lense–Thirring effect**: the precession of a gyroscope in Earth's gravitational field which is due to the rotation of the Earth. This is a smaller effect as compared to geodetic precession.

2.4.1 Geodetic Precession

Gyroscopes are rigid bodies which can also perform rotations described by S^{μ} besides the movement of its center of mass. To compute the geodetic precession of a gyroscope we use Eq. (2.92). In the local rest frame of the satellite (orbiting the Earth) we have for the spin vector:

$$S^{\prime\alpha} = (0, \vec{l}), \tag{2.94}$$

where $\vec{l}$ describes the angular momentum of the gyroscope. We use the standard form of the static and isotropic metric in spherical coordinates Eq. (1.26):

$$x^{\mu} = (ct, r, \theta, \varphi)$$
$$g = \text{diag}\big(B(r), -A(r), -r^2, -r^2\sin^2\theta\big).$$

We assume that the satellite is on a circular orbit, i.e.,

$$r = \text{const.}, \qquad \theta = \frac{\pi}{2}, \qquad \varphi = \omega_0\tau. \tag{2.95}$$

Therefore the velocity of the satellite is

$$u^{\mu} = \frac{dx^{\mu}}{d\tau} = \big(u^0 = \text{const.}, \ 0, \ 0, \ u^3 = \omega_0 = \text{const.}\big). \tag{2.96}$$

If we insert $\theta = \frac{\pi}{2}$ into the Christoffel symbols in Eq. (1.29), we obtain for the non-vanishing components

$$\Gamma^1{}_{00} = \frac{B'}{2A}, \quad \Gamma^1{}_{11} = \frac{A'}{2A}, \quad \Gamma^1{}_{22} = -\frac{r}{A}, \quad \Gamma^1{}_{33} = -\frac{r}{A}$$

$$\Gamma^0{}_{01} = \Gamma^0{}_{10} = \frac{B'}{2B}, \quad \Gamma^2{}_{12} = \Gamma^2{}_{21} = \frac{1}{r}, \quad \Gamma^3{}_{13} = \Gamma^3{}_{31} = \frac{1}{r}. \tag{2.97}$$

Equation (2.92) reads then

$$\frac{dS^0}{d\tau} = -\Gamma^0{}_{01}u^0 S^1 \tag{2.98}$$

$$\frac{dS^1}{d\tau} = -\Gamma^1{}_{00}u^0 S^0 - \Gamma^1{}_{33}u^3 S^3 \tag{2.99}$$

$$\frac{dS^2}{d\tau} = 0 \tag{2.100}$$

$$\frac{dS^3}{d\tau} = -\Gamma^3{}_{31}u^3 S^1. \tag{2.101}$$

We can immediately solve the third of these equations:

$$S^2(\tau) = \text{const.}, \tag{2.102}$$

which is the component of the spin (or angular momentum) of the gyroscope perpendicular to the satellite's orbit (θ-direction) and is constant.

Since $r = \text{const.}$, all coefficients in the system of linear differential equations (2.98)–(2.101) are constants. We differentiate Eq. (2.99) with respect to τ and insert (2.98) and (2.101) on the right-hand side. This yields

$$\frac{d^2 S^1}{d\tau^2} = \left[\Gamma^1{}_{00}\Gamma^0{}_{01}(u^0)^2 + \Gamma^1{}_{33}\Gamma^3{}_{31}(u^3)^2\right] S^1$$

$$\equiv -\omega^2 S^1. \tag{2.103}$$

With $(u^3)^2 = \omega_0^2$ and inserting the Christoffel symbols from (2.97), we get

$$\omega^2 = \omega_0^2 \left[-\frac{B'^2}{4AB}\left(\frac{u^0}{u^3}\right)^2 + \frac{1}{A}\right]. \tag{2.104}$$

In order to understand the ratio $\frac{u^0}{u^3}$ better, we look at the equation for the trajectory of the satellite (*geodesic equation*):

$$\frac{du^\mu}{d\tau} = -\Gamma^\mu{}_{\nu\lambda}u^\nu u^\lambda. \tag{2.105}$$

The $\mu = 1$ component of this equation is ($u^1 = 0$)

$$0 = \frac{du^1}{d\tau} = -\Gamma^1{}_{00}(u^0)^2 - \Gamma^1{}_{33}(u^3)^2, \tag{2.106}$$

from which we infer

$$\left(\frac{u^0}{u^3}\right)^2 = \frac{2r}{B'}. \tag{2.107}$$

Inserting this into (2.104), we find

$$\omega = \omega_0 \sqrt{\frac{1}{A}\left(1 - \frac{rB'}{2B}\right)}. \tag{2.108}$$

Using the Schwarzschild solution ($B = A^{-1} = 1 - \frac{2a}{r}$), this yields

$$\boxed{\omega = \omega_0\sqrt{1 - \frac{3a}{r}}}, \tag{2.109}$$

or in terms of the Robertson expansion

$$\omega = \omega_0\sqrt{1 - (1 + 2\gamma)\frac{a}{r}}. \tag{2.110}$$

Since $r = $ const., Eq. (2.103) becomes

$$\ddot{S}^1 + \omega^2 S^1 = 0, \tag{2.111}$$

which describes an harmonic oscillator. With initial conditions $S^1(0) = S$ and $\dot{S}^1(0) = 0$ the solution reads

$$S^1(\tau) = S\cos(\omega\tau), \qquad S^2 = \text{const.} \tag{2.112}$$

Inserting this into Eq. (2.101), we obtain by integration

$$S^3(\tau) = -\frac{S\omega_0}{r\omega}\sin(\omega\tau). \tag{2.113}$$

We proceed by studying the time dependence of the projection (S^1, S^3) (or (r, φ)-components) of the spin vector onto the orbital plane ($\theta = \frac{\pi}{2}$), see Fig. 2.3. The constant vector $\vec{e}_x$ can be written as

$$\vec{e}_x = \underbrace{\cos(\omega_0\tau)}_{=\varphi}\vec{e}_r - \underbrace{\sin(\omega_0\tau)}_{=\varphi}\vec{e}_\varphi. \tag{2.114}$$

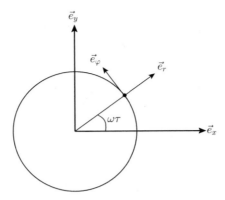

Fig. 2.3 Precession of the angular momentum vector (spin vector) takes place in the orbital plane of the satellite (described either by the (x, y)-plane or the (r, φ)-plane), since the θ-component of the spin vector (i.e., the component perpendicular to the orbital plane) is constant

The orbital period of the satellite is $\tau_0 = \frac{2\pi}{\omega_0}$. After each orbit, the argument φ in (2.114) increases by $\tau_0\omega_0 = 2\pi$ whereas the argument in (2.112) or (2.113) increases by $\tau_0\omega$ which differs slightly from 2π. The phase difference after one orbit is given by (note that in GR $\gamma = 1$)

$$\Delta\alpha = \tau_0(\omega_0 - \omega)$$
$$= 2\pi - 2\pi\sqrt{1 - \frac{(1 + 2\gamma)a}{r}}$$
$$\cong \pi\frac{(1 + 2\gamma)a}{r}. \tag{2.115}$$

Consider the vector $\vec{S}$ which is the projection of S^μ onto the orbital plane:

$$\vec{S} = S_r\vec{e}_r + S_\varphi\vec{e}_\varphi. \tag{2.116}$$

The components of this vector are given by

$$(S_r)^2 = -g_{11}S^1S^1 \quad \text{and} \quad (S_\varphi)^2 = -g_{33}S^3S^3 \tag{2.117}$$

with

$$-g_{11} = A(r) \quad \text{and} \quad -g_{33} = r^2\sin^2\theta\big|_{\theta=\frac{\pi}{2}} = r^2. \tag{2.118}$$

Therefore $S_r \propto \cos(\omega\tau)$ and $S_\varphi \propto \sin(\omega\tau)$. For $\tau = 0$, $\vec{S}$ is parallel to $\vec{e}_x$. However, after an orbit, $\omega\tau$ differs slightly from 2π as we have calculated in Eq. (2.115). The geodetic precession after one orbit is given by

$$\Delta\alpha = \frac{3\pi a}{r} \frac{(1+2\gamma)}{3}.$$ (2.119)

Let's consider a gyroscope on a satellite in a circular orbit around the Earth. We have (assuming $r \simeq R_E$)

$$\omega_0^2 R_E = \frac{GM_E}{R_E^2} = g \quad \text{and} \quad \tau_0 = 2\pi \left(\frac{R_E}{g}\right)^{1/2}.$$ (2.120)

After one year (i.e., after $\frac{t}{\tau_0}$ orbits with $t = 1$ year), we find

$$\Delta\alpha(t) = \Delta\alpha \frac{t}{\tau_0} = \frac{3\pi g R_E}{c^2} \frac{t}{2\pi\sqrt{\frac{R_E}{g}}} \simeq 8.4'' \, \text{yr}^{-1}.$$ (2.121)

For a generic radius r, one finds $\Delta\alpha(t) \simeq 8.4'' \, (R_E/r)^{5/2} \, \text{yr}^{-1}$.

On April 20, 2004, the satellite *Gravity Probe B* has been launched with the aim to measure the geodetic precession. At an altitude of 642 km, general relativity predicts a geodetic precession of -6606.1 mas yr^{-1} (1 mas = 1 milliarcsecond). The measured result -6601.8 ± 18.3 mas yr^{-1} is in excellent agreement with the prediction of GR [17, 18].

2.4.1.1 De Sitter Precession of the Moon

The Earth–Moon system can be considered as a "gyroscope" with an axis "perpendicular" to the plane containing Earth's orbit around the Sun. We denote by $\vec{l}$ the angular momentum of the Earth–Moon system with respect to the common center of mass ($\vec{l}$ is essentially the orbital angular momentum of the Moon because the common center of mass almost coincides with the center of the Earth).

The angular momentum $\vec{l}$ can be decomposed in a component $\vec{l}_\perp$ perpendicular to the Earth's orbital plane around the Sun and a parallel component $\vec{l}_\parallel$. The parallel component does not vanish because the orbital plane of the Moon around the Earth is tilted by 5° with respect to the orbital plane of the Earth around the Sun. In terms of the previously defined quantities, $\vec{l}_\perp$ corresponds to $\vec{S}^2$ and stays constant. The component $\vec{l}_\parallel$ corresponds to $\vec{S}$ and lies in the orbital plane of the Earth around the Sun. It is this parallel component which is affected by geodetic precession. Therefore, the orbital plane of the Moon rotates slightly (this was first noticed by De Sitter in 1916 [19]).

We can calculate the precession of Moon's orbital plane per century

$$\Delta\alpha_{\text{De Sitter}} = 100 \frac{3\pi a_\odot}{r_{\text{Earth-Sun}}} \simeq 2'' \text{per century},$$ (2.122)

where $a_\odot = 1.5$ km and $r_{\text{Earth-Sun}} = 1$ AU $\approx 150 \times 10^6$ km. Additionally we have a Newtonian precession (as in any three body system) with a period of 18.6 yr. This Newtonian effect is 10^7 times larger than the De Sitter precession. Nevertheless, the De Sitter precession has been measured. A laser beam has been sent to the Moon where it was reflected back to Earth by mirrors previously brought to the Moon by the Apollo missions (1969 and following years). The measurements that have been performed from 1970 till 1986 confirmed the De Sitter precession with a precision of about 1% [20, 21]. To discuss the Lense–Thirring effect we have to first derive the linearized field equations.

2.5 Linearized Field Equations

In order to find solutions to the Einstein field equations in the weak field regime, one can linearize the equations. These latter ones can then be used to calculate the Lense–Thirring effect and to describe gravitational waves.

Since the field equations are nonlinear, there is no standard procedure to solve the equations given a source for the fields $T_{\mu\nu}$. Besides numerical methods, there are essentially three ways to find solutions:

- **Exact solutions** assuming simplifying conditions (as, e.g., staticity, isotropy, ...). An example for this case is the Schwarzschild solution.
- Solutions of the **linearized field equations** for weak gravitational fields.
- Systematic expansion of the field equations and the equations of motion for weak fields and small velocities. This method is called **post-Newtonian approximation**. For example, in planetary systems we have $\frac{v^2}{c^2} \sim \frac{\phi}{c^2}$ which is very small. Such results should reproduce the Newtonian limit in lowest order (linearized in ϕ).

Here, we discuss the second possibility.

2.5.1 The Energy-Momentum Tensor of the Gravitational Field

The field itself is a form of energy and thus also a source of the field. This effect is of course due to the nonlinearities. For a weak field we can assume small deviations from the Minkowski metric:

$$g_{\mu\nu} = \eta_{\mu\nu} + h_{\mu\nu} \qquad \text{with } |h_{\mu\nu}| \ll 1. \tag{2.123}$$

One proceeds as follows: $G_{\mu\nu}$ has to be expanded in powers of $h_{\mu\nu}$ and the first order terms will lead to a linear wave equation. Terms of third order and higher are neglected, whereas the second-order terms describe the energy-momentum tensor of the gravitational field.

The expansion of the Ricci tensor can be written as

$$R_{\mu\nu} = R^{(1)}_{\mu\nu} + R^{(2)}_{\mu\nu} + \cdots, \tag{2.124}$$

since $R^{(0)}_{\mu\nu} = 0$. In order to compute the first-order term in (2.124), we write down the expansion of the curvature tensor[7]:

$$R_{\rho\mu\sigma\nu} = \frac{1}{2}\left(g_{\rho\sigma,\mu,\nu} + g_{\mu\nu,\rho,\sigma} - g_{\mu\sigma,\nu,\rho} - g_{\rho\nu,\sigma,\mu}\right) + \mathcal{O}(h^2), \tag{2.125}$$

where the derivatives are the usual non-covariant ones (the additional terms in the covariant derivatives are of higher order). We can thus write the first order Ricci tensor in terms of $h_{\mu\nu}$:

$$R^{(1)}_{\mu\nu} = \frac{1}{2}\left(\Box h_{\mu\nu} + h^{\rho}{}_{\rho,\mu,\nu} - h^{\rho}{}_{\mu,\rho,\nu} - h^{\rho}{}_{\nu,\rho,\mu}\right). \tag{2.126}$$

The d'Alembert operator can be used instead of $\partial_\mu \partial^\mu$ because in the approximation (2.123) the coordinates are "almost" Minkowskian, so $\partial_\mu \partial^\mu = \Box + \mathcal{O}(h)$. The first order Ricci scalar is given by

$$R^{(1)} = \eta^{\lambda\rho} R^{(1)}_{\lambda\rho}. \tag{2.127}$$

We proceed by considering the second-order equations. The left-hand side of the field equations can be written in terms of the quantity $t_{\mu\nu}$ which is defined by

$$R^{(2)}_{\mu\nu} - \left(\frac{R g_{\mu\nu}}{2}\right)^{(2)} =: \frac{8\pi G}{c^4} t_{\mu\nu}. \tag{2.128}$$

We take these terms to the right-hand side of Einstein's equations and find at second order in $h_{\mu\nu}$:

$$R^{(1)}_{\mu\nu} - \frac{R^{(1)}}{2}\eta_{\mu\nu} = -\frac{8\pi G}{c^4}(T_{\mu\nu} + t_{\mu\nu}). \tag{2.129}$$

This can be interpreted as a wave equation linear in $h_{\mu\nu}$ with source terms

$$\tau_{\mu\nu} = T_{\mu\nu} + t_{\mu\nu}. \tag{2.130}$$

We have to consider $\tau_{\mu\nu}$ as being the energy-momentum tensor which also includes the contribution of the gravitational field itself.

---

[7] Notice that we adopt here the "minus" convention for the Einstein field equations (i.e., $R_{\mu\nu} - \frac{R}{2}g_{\mu\nu} = -\frac{8\pi G}{c^4}T_{\mu\nu}$).

We interpret (2.129) as follows: due to $G^{\mu\nu}{}_{;\nu} = 0$ (Bianchi identity), the left-hand side of (2.129) has to satisfy (to the considered order in h):

$$\frac{\partial}{\partial x_\nu}\left(R^{(1)}_{\mu\nu} - \frac{R^{(1)}}{2}\eta_{\mu\nu}\right) = 0, \tag{2.131}$$

and, therefore, for the right-hand side we get

$$\frac{\partial \tau_{\mu\nu}}{\partial x_\nu} = 0. \tag{2.132}$$

This gives the momentum

$$P_\mu = \int d^3r \; \tau_{\mu 0} = \text{const.} \tag{2.133}$$

which is conserved (in time). We can thus interpret $\tau_{\mu 0}$ as the momentum density and $\tau_{\mu\nu}$ as an energy-momentum tensor. Since $T_{\mu\nu}$ includes all non-gravitational sources and $\tau_{\mu\nu}$ can be viewed as the "total" energy-momentum tensor, $t_{\mu\nu}$ describes thus the energy-momentum which is only due to the gravitational field:

$$t^{\text{grav.}}_{\mu\nu} = \frac{c^4}{8\pi G}\left(R^{(2)}_{\mu\nu} - \left(\frac{Rg_{\mu\nu}}{2}\right)^{(2)}\right) \qquad (|h_{\mu\nu}| \ll 1). \tag{2.134}$$

2.5.2 Linearized Field Equations

With Eq. (2.125) we find for the field equations at first order in h

$$\Box h_{\mu\nu} + h^\rho{}_{\rho,\mu,\nu} - h^\rho{}_{\mu,\rho,\nu} - h^\rho{}_{\nu,\rho,\mu} = -\frac{16\pi G}{c^4}\left(T_{\mu\nu} - \frac{T}{2}\eta_{\mu\nu}\right). \tag{2.135}$$

We use $\eta_{\mu\nu}$ instead of $g_{\mu\nu}$ in this equation because both sides are already of order h. Since the field equations are covariant, we are free to perform a coordinate transformation. Note, however, that since $|h_{\mu\nu}| \ll 1$ we can perform coordinate transformations which deviate only slightly from Minkowski coordinates:

$$x^\mu \longrightarrow x'^\mu = x^\mu + \varepsilon^\mu(x) \qquad \text{with } \varepsilon \ll 1. \tag{2.136}$$

From $g'^{\mu\nu} = \frac{\partial x'^\mu}{\partial x^\lambda}\frac{\partial x'^\nu}{\partial x^\rho} g^{\lambda\rho}$ we infer how $h_{\mu\nu}$ transforms. With $\frac{\partial x'^\mu}{\partial x^\lambda} = \delta^\mu_\lambda + \frac{\partial \varepsilon^\mu}{\partial x^\lambda}$ inserted into $g'^{\mu\nu}$ we get

$$g'^{\mu\nu} = \eta^{\mu\nu} - h'^{\mu\nu}$$

$$= \left(\delta_\lambda^\mu + \frac{\partial\varepsilon^\mu}{\partial x^\lambda}\right)\left(\delta_\rho^\nu + \frac{\partial\varepsilon^\nu}{\partial x^\rho}\right)\left(\eta^{\lambda\rho} - h^{\lambda\rho}\right), \tag{2.137}$$

where we used that from $g_{\mu\nu} = \eta_{\mu\nu} + h_{\mu\nu}$ it follows $g^{\mu\nu} = \eta^{\mu\nu} - h^{\mu\nu}$. From Eq. (2.137) we infer

$$h'^{\mu\nu} = h^{\mu\nu} - \frac{\partial\varepsilon^\mu}{\partial x_\nu} - \frac{\partial\varepsilon^\nu}{\partial x_\mu}. \tag{2.138}$$

Since this is already a first-order equation (in h), we can raise and lower indices with $g_{\mu\nu} \simeq \eta_{\mu\nu}$ and $g^{\mu\nu} \simeq \eta^{\mu\nu}$. Thus

$$h'_{\mu\nu} = h_{\mu\nu} - \frac{\partial\varepsilon_\mu}{\partial x^\nu} - \frac{\partial\varepsilon_\nu}{\partial x^\mu}. \tag{2.139}$$

In analogy to electrodynamics this transformation of the "potentials" $g_{\mu\nu}$ is called a **gauge transformation**. We can choose four functions $\varepsilon^\mu(x)$ which give four constraints on the "potentials" $h_{\mu\nu}$. For instance,

$$2h^\mu{}_{\nu,\mu} = h^\mu{}_{\mu,\nu}. \tag{2.140}$$

We insert the gauge condition (2.140) into (2.135) and obtain the **decoupled linearized field equations**:

$$\boxed{\Box h_{\mu\nu} = -\frac{16\pi G}{c^4}\left(T_{\mu\nu} - \frac{T}{2}\eta_{\mu\nu}\right),} \tag{2.141}$$

as can easily be seen if we differentiate (2.140) (i.e., $h^\rho{}_{\rho,\mu} = 2h^\rho{}_{\mu,\rho}$) with respect to x^ν:

$$h^\rho{}_{\rho,\mu,\nu} = 2h^\rho{}_{\mu,\rho,\nu} = h^\rho{}_{\mu,\rho,\nu} + h^\rho{}_{\nu,\rho,\mu} \tag{2.142}$$

(we used $h_{\mu\nu} = h_{\nu\mu}$). This implies

$$-h^\rho{}_{\mu,\rho,\nu} - h^\rho{}_{\nu,\rho,\mu} + h^\rho{}_{\rho,\mu,\nu} = 0, \tag{2.143}$$

which is just another form of our gauge condition from which it can be seen that (2.135) indeed reduces to (2.141).

Furthermore, it can be shown that from (2.139) it follows that if $h_{\mu\nu}$ does not satisfy (2.140), then we can find a transformed $h'_{\mu\nu}$ that does so. This can be done using the coordinate transformation (2.136) with $\Box\varepsilon_\nu = h^\mu{}_{\nu,\mu} - \frac{1}{2}h^\mu{}_{\mu,\nu}$.

The linearized field equation Eq. (2.141) has the same structure as the field equations in electrodynamics. We can therefore immediately write down the well-known solution for the retarded potentials:

$$
h_{\mu\nu}(\vec{r}, t) = -\frac{4G}{c^4} \int d^3 r' \, \frac{S_{\mu\nu}\left(\vec{r}', t - \frac{|\vec{r}-\vec{r}'|}{c}\right)}{|\vec{r} - \vec{r}'|} \tag{2.144}
$$

$$
\text{with} \qquad S_{\mu\nu} = T_{\mu\nu} - \frac{T}{2}\eta_{\mu\nu}.
$$

The interpretation is the same as in electrodynamics: a change in $S_{\mu\nu}$ at position $\vec{r}'$ does not have any influence at the position $\vec{r}$ before the time $\frac{|\vec{r}-\vec{r}'|}{c}$ has passed.

2.6   Lense–Thirring Effect

The classical example of the Lense–Thirring effect is the precession of a gyroscope in Earth's gravitational field due to the Earth's rotation. To set up an analogy with electrodynamics, we note that the gravitational field of the Schwarzschild metric would correspond to the Coulomb field outside of a static, spherical charge distribution. If the charge distribution rotates with constant angular velocity, there will be a static, non-isotropic magnetic field. Similarly, the rotation of the Earth will cause a *gravitomagnetic* field.

We will treat this problem by using the linearized field equations as derived in the previous section. Another approach would be to start from the Kerr solution (i.e., the metric outside of a rotating black hole) and apply the weak field limit.

2.6.1   Metric of the Rotating Earth

We assume a weak field caused by a slowly rotating planet, so $|h_{\mu\nu}| \ll 1$. The linearized field equations read

$$
\Box h_{\mu\nu} = -\frac{16\pi G}{c^4} \left(T_{\mu\nu} - \frac{T}{2}\eta_{\mu\nu}\right). \tag{2.145}
$$

The coordinates $x^\mu = (x^0, \ldots, x^3)$ are Minkowski coordinates up to corrections of $\mathcal{O}(h)$. In the energy-momentum tensor Eq. (2.42) we can neglect the pressure since $p \ll \rho c^2$. Since the velocities (rotation of the Earth) are small compared to c, we neglect terms of order $\left(\frac{v}{c}\right)^2$. The energy-momentum tensor thus reads

$$T_{\mu\nu} \simeq \rho c^2 \begin{pmatrix} 1 & \frac{v_i}{c} \\ \frac{v_i}{c} & 0 \end{pmatrix} \quad (i = 1, 2, 3). \tag{2.146}$$

The terms proportional to v_i generate the gravitomagnetic field. This has to be compared to electrodynamics, where magnetic fields are generated by currents.

The mass distribution of the Earth can be approximated as follows:

$$\rho(\vec{r}) = \begin{cases} \rho_0 & (r \le R_E) \\ 0 & (r > R_E). \end{cases} \tag{2.147}$$

The angular velocity of the Earth is

$$\vec{\omega} = \omega \vec{e}_3 \quad \text{with} \quad \omega = \frac{2\pi}{1\,\text{day}}. \tag{2.148}$$

We consider the Earth as a rigid body, so we can write the velocity field as

$$\vec{v}(\vec{r}) = \vec{\omega} \wedge \vec{r} \quad \text{or} \quad v_i = \varepsilon_{ijk}\omega^j x^k. \tag{2.149}$$

Since this velocity is constant, $T_{\mu\nu}$ does not depend on time and therefore the field equations (2.145) have stationary solutions. We can thus replace $\Box$ by $-\Delta$ and Eq. (2.145) becomes

$$\Delta h_{\mu\mu} = \frac{8\pi G}{c^2}\rho(\vec{r}), \tag{2.150}$$

$$\Delta h_{0i} = \frac{16\pi G}{c^3}\rho(\vec{r})\varepsilon_{ijk}\omega^j x^k, \tag{2.151}$$

where we used $\eta_{\mu\nu} = \text{diag}(1, -1, -1, -1)$, $\frac{T}{2} = \frac{\rho c^2}{2}$ and $T_{00} = \rho c^2$, $T_{ii} = 0$.
  Using

$$\Delta \frac{1}{|\vec{r} - \vec{r}'|} = -4\pi \delta^{(3)}(\vec{r} - \vec{r}'), \tag{2.152}$$

we can immediately solve Eqs. (2.150) and (2.151):

$$h_{\mu\mu}(\vec{r}) = -\frac{2G}{c^2} \int d^3 r' \frac{\rho(\vec{r}')}{|\vec{r} - \vec{r}'|}, \tag{2.153}$$

$$h_{0i}(\vec{r}) = -\frac{4G}{c^3}\varepsilon_{ikn}\omega^k \int d^3 r' \frac{\rho(\vec{r}')x'^n}{|\vec{r} - \vec{r}'|}, \tag{2.154}$$

where x'^n denotes the nth component of $\vec{r}'$.
  For the region outside of the mass distribution ($r > R_E$), we can use the expansion

$$\frac{1}{|\vec{r} - \vec{r}'|} = \sum_{l,m} \frac{4\pi}{(2l+1)} \frac{r'^l}{r^{l+1}} Y_{lm}^*(\hat{\vec{r}}') Y_{lm}(\hat{\vec{r}}) \qquad (r > r')$$

$$= \frac{1}{r} - \frac{x^j x_j'}{r^3} + \dots , \qquad (2.155)$$

where Y_{lm} are the spherical harmonic functions and $\hat{\vec{r}}$, $\hat{\vec{r}}'$ denote unit vectors in directions of $\vec{r}$ and $\vec{r}'$, respectively. Note that for the Cartesian components we have $x_i = g_{ik} x^k = -x^i + \mathcal{O}(h)$. The corrections of $\mathcal{O}(h)$ can thus be neglected because the right-hand sides of (2.153) and (2.154) are already of first order in h. Since $\rho(\vec{r})$ is spherically symmetric, only the first term of (2.155) contributes in Eq. (2.153):

$$h_{\mu\mu}(\vec{r}) = -\frac{2G}{c^2 r} \int d^3 r' \, \rho(\vec{r}') = -\frac{2G M_E}{c^2 r} \quad (r \ge R_E). \qquad (2.156)$$

Since $\rho x'^n$ is proportional to Y_{1m}, only terms with $l = 1$ contribute to (2.154):

$$h_{0i}(\vec{r}) = \frac{4G}{c^3} \varepsilon_{ijn} \frac{\omega^j x^k}{r^3} \int d^3 r' \, \rho(\vec{r}') x'^n x_k'$$

$$= -\frac{4G M_E R_E^2}{5c^3} \varepsilon_{ijn} \frac{\omega^j x^n}{r^3} \qquad (r \ge R_E). \qquad (2.157)$$

Since $\rho(\vec{r})$ is spherically symmetric, we integrated using $x'^n x_k' = -\delta_k^n \frac{r'^2}{3}$ and furthermore used $\rho_0 = 3M_E/(4\pi R_E^3)$. Considering h_{0i} as a vector, i.e., $h_{0i} \to \vec{h} = h_{0i}\vec{e}^i$, we can write (2.151) as

$$\Delta \vec{h}(\vec{r}) = \frac{16\pi G}{c^3} \rho \vec{\omega} \wedge \vec{r} \qquad (2.158)$$

and (2.157) becomes

$$\boxed{\vec{h}(\vec{r}) = -\frac{4G M_E R_E^2}{5c^3} \frac{\vec{\omega} \wedge \vec{r}}{r^3} = -\frac{2G I}{c^3} \frac{\vec{\omega} \wedge \vec{r}}{r^3},} \qquad (2.159)$$

where $I = \frac{2}{5} M_E R_E^2$ is the moment of inertia of a homogeneous sphere.

Let's consider the analogy with electrodynamics again. In magnetostatics, the vector potential $\vec{A}$ of a homogeneously charged rotating sphere with radius R and total charge q satisfies

$$\Delta \vec{A} = -\frac{4\pi}{c} \rho_e \vec{\omega} \wedge \vec{r} \qquad (2.160)$$

$$\Rightarrow \quad \vec{A} = \frac{q R^2}{5c} \frac{\vec{\omega} \wedge \vec{r}}{r^3}, \qquad (2.161)$$

which has the same form as Eq. (2.159).

Equations (2.156) and (2.157) determine the metric of the rotating Earth (valid for $r \geq R_E$):

$$ds^2 = \left(1 - \frac{2GM_E}{c^2 r}\right) c^2 dt^2 - \left(1 + \frac{2GM_E}{c^2 r}\right) d\vec{r}^2 + 2c h_{0i}\, dx^i dt, \qquad (2.162)$$

where $d\vec{r}^2 = -dx^i dx_i$. Note that this metric at $\mathcal{O}\left(\frac{GM_E}{c^2 r}\right)$ and for $\omega = 0$ does not reduce to the Schwarzschild metric since Eq. (2.145) implies that we chose other coordinates as compared to the standard form. However, the metric (2.162) asymptotically ($r \to \infty$) becomes the Minkowski metric. One can perform a coordinate transformation such that $d\vec{r}^2$ has the usual angular dependence, i.e., $d\vec{r}^2 \to dr^2 + r^2(d\theta^2 + \sin^2\theta d\varphi^2)$. Distant "fixed" stars (which live in the asymptotic Minkowski spacetime) have constant values for the angles (θ, φ). Therefore changes in the angles due to spin precession refer to rotations with respect to the distant fixed stars.

2.6.1.1 Rotation of the Local IS

The metric (2.162) implies a rotation of the local inertial system IS. In order to see this, consider the axis of a gyroscope which is described by (see 2.92)

$$\frac{dS^\mu}{d\tau} = -\Gamma^\mu{}_{\kappa\nu} S^\kappa u^\nu. \qquad (2.163)$$

A freely falling gyroscope would not only show a precession linked to the Lense–Thirring effect (that can be loosely defined as "precession due to the rotation of the central body"), but would also be subject, because of its motion in a curved metric, to the geodesic precession derived in Sect. 2.4. However, we want here to "separate" these two different contributions, and only look at the Lense–Thirring component. This can be done with a simple trick: We consider the gyroscope as being at rest in the external frame, and just let it free to change its rotational axis. Thus we have, in the coordinate system defined by Eq. (2.162),

$$S^\mu = (0, S^i), \qquad u^\mu = (c, \vec{0}). \qquad (2.164)$$

Because we are taking into account only terms up to $\mathcal{O}(h)$, we can replace $\frac{dS^i}{d\tau}$ by $\frac{dS^i}{dt}$. From Eqs. (2.163) and (2.164) it follows that

$$\frac{dS^i}{dt} = -c\Gamma^i{}_{0j} S^j. \qquad (2.165)$$

For time independent $h_{\mu\nu}$ we find at first order

$$\Gamma^i{}_{0j} = \frac{\eta^{ik}}{2} \left(\frac{\partial h_{0k}}{\partial x^j} - \frac{\partial h_{0j}}{\partial x^k} \right) = \frac{1}{2} \left(\partial_j h_0{}^i - \partial^i h_{0j} \right). \tag{2.166}$$

Inserting this into (2.165) and lowering the index i, we find

$$\begin{aligned}
\frac{dS_i}{dt} &= -\frac{c}{2} \left(\partial_j h_{0i} - \partial_i h_{0j} \right) S^j \\
&= -\frac{c}{2} \varepsilon_{ikl} \left(\varepsilon^{kmn} \partial_m h_{0n} \right) S^l \\
&= \varepsilon_{ikl} \Omega^k S^l
\end{aligned} \tag{2.167}$$

with $\varepsilon_{ikl}\varepsilon^{kmn} = \delta_l^m \delta_i^n - \delta_l^n \delta_i^m$ and the **gravitomagnetic field**

$$\Omega^k = -\frac{c}{2} \varepsilon^{kmn} \partial_m h_{0n} \quad \text{or} \quad \vec{\Omega}(\vec{r}) = -\frac{c}{2} \vec{\nabla} \wedge \vec{h}(\vec{r}). \tag{2.168}$$

The indices i, j, m run over 1, 2, 3. We call $\vec{h}$ the **gravitomagnetic potential**. Using $\vec{h}$ from Eq. (2.159) we obtain the following expression for the angular velocity in the local IS:

$$\boxed{\vec{\Omega}(\vec{r}) = \frac{2G M_E R_E^2}{5c^2} \frac{3(\vec{\omega} \cdot \vec{r})\vec{r} - \vec{\omega}r^2}{r^5}.} \tag{2.169}$$

Note that $\vec{\Omega}$ has the same form as $\vec{B} = \vec{\nabla} \wedge \vec{A}$ in electrodynamics. In vector form, Eq. (2.167) reads

$$\boxed{\frac{d\vec{S}}{dt} = \vec{\Omega} \wedge \vec{S} \quad \text{or} \quad d\vec{S} = (\vec{\Omega}dt) \wedge \vec{S}.} \tag{2.170}$$

This implies a precession of the spin of the gyroscope's axis with angular velocity $\vec{\Omega}$. This precession is due to the rotation of the Earth with angular velocity $\vec{\omega}$. The precession of the gyroscope's axis is equivalent to the rotation of the local IS because in the local IS we have $\vec{S} = $ const. Therefore, the local IS rotates with angular velocity $\vec{\Omega}$ as compared to the global coordinate system described by (2.162). For $r \to \infty$ Eq. (2.162) becomes Minkowskian, i.e., it describes a system which doesn't rotate with respect to the fixed star system.

To summarize, the physical meaning of the angular velocity $\vec{\Omega}$ is that the local IS rotates with Ω with respect to the fixed star system. The rotation of the Earth drags the local IS ("**frame dragging**").

Now that we derived geodetic precession and the Lense–Thirring effect, we note that in fact both effects take place at the same time and thus sum up. Inserting $r = R_E$ in Eq. (2.169) we get at the north pole and at the equator, respectively:

$$\vec{\Omega} = \frac{2GM_E}{5c^2 R_E} \vec{\omega} \cdot \begin{cases} 2 & (\text{North pole}, \theta = 0) \\ (-1) & (\text{equator}, \theta = \frac{\pi}{2}), \end{cases} \tag{2.171}$$

where we used $|\vec{\omega} \cdot \vec{r}| = \omega r \cos \theta$. This evaluates to

$$\frac{2GM_E}{5c^2 R_E} \omega \sim 10^{-9} \omega. \tag{2.172}$$

We conclude that the Lense–Thirring precession affects a Foucault pendulum located at the north pole. The rotation with respect to the distant stars amounts to

$$\Delta\phi = \Omega_{LT} \cdot 1\text{yr} = \frac{4GM_E}{5c^2 R_E} 2\pi \cdot 365 = 0.2'' \text{ per year.} \tag{2.173}$$

This effect was first computed by Lense and Thirring in 1918 [22].

If one chooses the spin perpendicular to the orbital plane, the Lense–Thirring effect vanishes for equatorial orbits since in this case we have $\vec{S}||\vec{w}||\vec{\Omega}$ and Eq. (2.170) gives $\frac{d\vec{S}}{dt} = 0$, whereas would be maximal for polar orbits. The expected total frame dragging is at most of the order of $0.05''$ per year. The NASA satellite *Gravity Probe B*, with the aim to measure both geodetic and Lense–Thirring precession, was launched in 2004 in a polar orbit around Earth. In May 2011 the final results of *Gravity Probe B* have been released, and they indeed confirm the GR predictions for the Lense–Thirring effect . The measured value being: -37.2 ± 7.2 mas/yr (1 mas = 1 milliarcsecond) to be compared with the GR prediction of -39.2 mas/yr [23]. However, there are claims that the effect has been measured on the orbit of the LAGEOS satellites as well [24].

2.6.2 Gravitomagnetic Forces

In the discussion of the Lense–Thirring precession we saw a formal equivalence to electromagnetism:

$$\vec{h} \quad \leftrightarrow \quad \vec{A}$$
$$\vec{\Omega} \quad \leftrightarrow \quad \vec{B}.$$

This analogy persists even for the equation of motion of a particle in the metric (2.162),

$$\frac{du^\mu}{d\tau} = -\Gamma^\mu{}_{\gamma\nu} u^\gamma u^\nu. \tag{2.174}$$

Neglecting terms $\mathcal{O}\left(v^2/c^2\right)$ we have $\mathrm{d}\tau \approx \mathrm{d}t$ and $u^\mu = (c, v^i)$. Therefore Eq. (2.174) reads

$$\frac{\mathrm{d}v^i}{\mathrm{d}t} = -\Gamma^i{}_{00}c^2 - 2c\Gamma^i{}_{0j}v^j + \mathcal{O}\left(\frac{v^2}{c^2}\right). \tag{2.175}$$

The first term on the right-hand side corresponds to the gradient of the Newtonian potential ϕ. In analogy to to (2.165)–(2.167), the second term can be shown to give

$$-\Gamma^i{}_{0j}v^j = (\vec{\Omega} \wedge \vec{v})^i. \tag{2.176}$$

This implies that Eq. (2.175) can be written as

$$\frac{\mathrm{d}\vec{v}}{\mathrm{d}t} = -\mathrm{grad}\phi + 2\vec{\Omega} \wedge \vec{v}. \tag{2.177}$$

This is the equation of motion in the presence of gravitomagnetic forces, and it has the same structure as the equation for the Lorentz force

$$\vec{K} = q\left(\vec{E} + \frac{1}{c}\vec{v} \wedge \vec{B}\right). \tag{2.178}$$

This analogy is the origin of the notion of **gravitomagnetism**.

Note, however, that this analogy is only true if we consider the linearized field equations of general relativity. The electromagnetic fields are exact solutions to Maxwell's equations, whereas the gravitomagnetic potential and field are approximations. Furthermore, the analogy is quite formal and certainly not complete due to the absence of negative "gravitational charges".

2.7 Summary

Gravitational lensing is nowadays an important tool to study the universe: from our galaxy using microlensing, to distant galaxies and galaxy clusters. Gravitational lensing is a way to learn about the matter and dark matter content at various scales as well as to determine cosmological parameters, like the Hubble constant. We discussed the main tools used in gravitational lensing. To describe neutron stars one needs to take into account general relativistic effects as discussed in Sect. 2.2. The study of neutron stars will undoubtedly progress in future thanks to more observations in the electromagnetic bands and with gravitational waves as was the case on August 17, 2017, when the LIGO-Virgo detectors observed the gravitational wave signal coming from the coalescence of two neutron stars in the Galaxy NGC 4993 at a distance of about 130 million light years from Earth. The gravitational wave observation was then followed by the detections in many electromagnetic bands. Such observations will help to better understand neutron stars, especially the equation of state of the matter in its interior. The time delay effect, as proposed in 1964 by Shapiro, is by now a classical tool to test general relativity in our solar system, but it is also of relevance in the study of binary pulsar systems. Geodetic precession and the Lense–Thirring effect have been verified by the Gravity Probe B satellite, which was put 2004 into a polar orbit around Earth. Again these effects are of relevance in the accurate study of the measurements of binary pulsar systems. To derive the Lense–Thirring effect we discussed the linearized field equations. These equations will be used in the next chapter to discuss the gravitational waves.

2.8 Problems

Problem 2.1 *(Metric of a static star)*

The exterior spacetime of extended spherical objects can be approximated by the Schwarzschild metric. In this exercise we will derive the interior metric modeling the star by a perfect fluid with energy-momentum tensor $T_{\mu\nu} = (\rho + p)u_\mu u_\nu - p g_{\mu\nu}$. As a first step we will consider the general static, spherically symmetric metric

$$ds^2 = \exp[2\alpha(r)]dt^2 - \exp[2\beta(r)]dr^2 - r^2 d\Omega^2. \tag{2.179}$$

The corresponding non-vanishing Christoffel symbols are (up to permutations of the two lower indices)

$$\Gamma^t_{tr} = \partial_r\alpha \qquad \Gamma^r_{tt} = \partial_r\alpha \exp[2(\alpha - \beta)]$$
$$\Gamma^r_{rr} = \partial_r\beta \qquad \Gamma^r_{\theta\theta} = -r\exp[-2\beta] \qquad \Gamma^r_{\phi\phi} = -r\sin^2(\theta)\exp[-2\beta]$$
$$\Gamma^\theta_{\theta r} = r^{-1} \qquad \Gamma^\theta_{\phi\phi} = -\sin(\theta)\cos(\theta)$$
$$\Gamma^\phi_{\phi r} = r^{-1} \qquad \Gamma^\phi_{\phi\theta} = \cot(\theta),$$

while the Ricci tensor takes the form

$$R_{tt} = \exp[2(\alpha - \beta)]\left(\partial_r^2\alpha + (\partial_r\alpha)^2 - \partial_r\alpha\partial_r\beta + \frac{2}{r}\partial_r\alpha\right) \tag{2.180}$$

$$R_{rr} = -\partial_r^2\alpha - (\partial_r\alpha)^2 + \partial_r\alpha\partial_r\beta + \frac{2}{r}\partial_r\beta \tag{2.181}$$

$$R_{\theta\theta} = 1 - \exp[-2\beta] + (\partial_r\beta - \partial_r\alpha)\,r\exp[-2\beta] \tag{2.182}$$

$$R_{\phi\phi} = \sin^2(\theta)R_{\theta\theta}. \tag{2.183}$$

(a) In the fluid rest frame the velocity is pointing in the timelike direction. Using this and the normalization $u_\mu u^\mu = 1$ show that the energy-momentum tensor can be written as

$$T_{\mu\nu} = \text{diag} \left\{ \exp\left[2\alpha\right] \rho, \exp\left[2\beta\right] p, r^2 p, r^2 \sin^2(\theta) p \right\}. \tag{2.184}$$

(b) Write down the Einstein equations inside the star.
It will be convenient to define

$$m(r) = \frac{1}{2G}\left(r - r\exp\left[-2\beta\right]\right) \tag{2.185}$$

and to rewrite the Einstein equations in terms of $m(r)$.

(c) Use the energy-momentum conservation $\nabla_\mu T^{\mu\nu} = 0$ to show

$$(\rho + p)\frac{d\alpha}{dr} = -\frac{dp}{dr}. \tag{2.186}$$

Use the above equation together with the rr-component of the Einstein equations to derive the Tolman–Oppenheimer–Volkoff equation of a star

$$\frac{dp}{dr} = -\frac{(\rho + p)\left[Gm(r) + 4\pi G r^3 p\right]}{r\left[r - 2Gm(r)\right]}. \tag{2.187}$$

This is the relativistic version of the equation of hydrostatic equilibrium for a star.

(d) Assuming an incompressible fluid with constant density ρ_* out to the surface of the star, solve for the pressure $p(r)$ and write down the metric. What is the maximum mass a star of a given radius R can have?

Problem 2.2 *(Linearized field equations)*

In linearized gravity we consider the metric to be of the form $g_{\mu\nu} = \eta_{\mu\nu} + h_{\mu\nu}$ where $|h_{\mu\nu}| \ll 1$. This allows us to drop all terms $\mathcal{O}((h_{\mu\nu})^2)$. Furthermore, we drop all derivatives $\mathcal{O}((\partial_\rho h_{\mu\nu})^2)$.

(a) Show that the Einstein field equations reduce to the linearized field equations

$$\Box h_{\mu\nu} + h_{,\mu,\nu} - h^\rho{}_{\mu,\rho,\nu} - h^\rho{}_{\nu,\rho,\mu} = -16\pi G \left(T_{\mu\nu} - \frac{1}{2}T\eta_{\mu\nu} \right), \tag{2.188}$$

where $\Box = \partial^\rho \partial_\rho$ is the d'Alembert operator and $h = h^\mu{}_\mu$ is the trace of the perturbation, by following these steps:

(i) Show that the inverse metric is given by $g^{\mu\nu} = \eta^{\mu\nu} - h^{\mu\nu}$ and that $h = \eta^{\mu\nu}h_{\mu\nu}$.

(ii) Calculate the Ricci tensor: $R_{\mu\nu} = R^\rho{}_{\mu\rho\nu} = \partial_\rho \Gamma^\rho{}_{\nu\mu} - \partial_\nu \Gamma^\rho{}_{\rho\mu} + \Gamma^\rho{}_{\rho\lambda}\Gamma^\lambda{}_{\nu\mu} - \Gamma^\rho{}_{\nu\lambda}\Gamma^\lambda{}_{\rho\mu}$.

(iii) Use the Einstein field equations: $R_{\mu\nu} = 8\pi G \left(T_{\mu\nu} - \frac{1}{2}T g_{\mu\nu}\right)$.

(b) (i) Consider a change of coordinates $\tilde{x}^\mu = x^\mu - \xi^\mu(x)$, with $\xi^\mu(x)$ being a vector field and $|\partial_\alpha \xi^\mu| \ll 1$. Show that under this change of coordinates, $h_{\mu\nu}$ changes to

$$\tilde{h}_{\mu\nu} = h_{\mu\nu} + (\partial_\mu \xi_\nu + \partial_\nu \xi_\mu), \qquad (2.189)$$

where all terms of order $\mathcal{O}((\partial\xi)^2)$ have been dropped. This is called a "gauge transformation": By choosing the four functions ξ_μ, we obtain four constraints on $h_{\mu\nu}$.

(ii) Show that, choosing the gauge $h_{\mu\nu}{}^{,\mu} = \frac{1}{2} h_{,\nu}$, the field equations read

$$\Box h_{\mu\nu} = -16\pi G \left(T_{\mu\nu} - \frac{1}{2}T \eta_{\mu\nu}\right). \qquad (2.190)$$

(iii) Show that for a static, general source described by $T_{\mu\nu} = \text{diag}\{\rho, 0, 0, 0\}$, the metric $g_{\mu\nu}$ reduces to:

$$ds^2 = (1 + 2\phi)\, dt^2 - (1 - 2\phi)\left(dx^2 + dy^2 + dz^2\right), \qquad (2.191)$$

where ϕ is the Newtonian gravitational potential associated to ρ.

(c) Consider the trace reversed form of the metric perturbations,

$$\gamma_{\mu\nu} = h_{\mu\nu} - \frac{1}{2}\eta_{\mu\nu} h. \qquad (2.192)$$

Show that in the Lorenz gauge (in which the divergence of $\gamma_{\mu\nu}$ vanishes: $\gamma_{\mu\nu}{}^{,\nu} = 0$), the linearized field equations become

$$\Box \gamma_{\mu\nu} = -16\pi G T_{\mu\nu}. \qquad (2.193)$$

Problem 2.3 *(Out of plane precession of S2 orbit)*

The center of our galaxy harbors a rotating black hole. Many of the stars within 1 pc from the central black hole have orbits that lie in a plane. This suggests that these stars formed from gas that is arranged in a disk-like structure. However, the orbit of the star S2 with semi-major axis $a_r = 4.7 \times 10^{-3}$ pc and eccentricity $e = 0.88$ is inclined by $75°$ with respect to this plane. One possible origin for this inclination could be the precession of the orbital plane due to the black hole's spin. Using the age estimate of $t_{S2} \approx 3 \times 10^6$ yr, calculate a lower bound on the black hole spin using the black hole mass $M_{BH} = 4.2 \times 10^6\, M_\odot$.

Use the "gravitomagnetic" equations of motion (see Sect. 2.6.2) written as

$$\frac{d\vec{v}}{dt} = -\nabla\phi + 2\vec{\Omega} \wedge \vec{v}, \tag{2.194}$$

where

$$\vec{\Omega} = \frac{G}{c^2}\left[\frac{3\vec{r}(\vec{r}\cdot\vec{S})}{r^5} - \frac{\vec{S}}{r^3}\right]. \tag{2.195}$$

Here, $\vec{S}$ is the spin of the black hole. Since a lower bound for the spin is wanted, it is sufficient to estimate the precession rate for the configuration where the precession is maximal, i.e., when $\vec{S}$ lies on the orbital plane. Furthermore, simplify the system assuming a corresponding circular orbit. Note that the black hole spin is conveniently parametrized in terms of the spin parameter

$$a = \frac{Sc}{GM^2}, \tag{2.196}$$

where $a < 1$. A black hole with $a = 1$ is said to be maximally spinning.

Problem 2.4 *(Gravitational field of a moving particle)*

Consider a particle of mass M moving with constant velocity $\vec{V}$. Calculate the gravitomagnetic potential and the equation of motion for a test particle to order $\mathcal{O}(v/c)$.

Start in the particles rest frame Σ' with Eqs. (2.153) and (2.154) for the metric perturbation and then perform a coordinate transformation to the global frame Σ. Since we are working at $\mathcal{O}(h)$ the transformation is nothing than a Lorentz boost.

References

1. J.G. von Soldner, *Ueber die Ablenkung eines Lichtstrahls von seiner geradlinigen Bewegung, durch die Attraktion eines Weltkörpers, an welchem er nahe vorbei geht* (Berliner Astronomisches Jahrbuch, 1804), S. 161–172
2. A. Einstein, Science **84**, 506 (1936)
3. F. Zwicky, Phys. Rev. **51**, 290 (1937)
4. F. Zwicky, Phys. Rev. **51**, 678 (1937)
5. P. Schneider, J. Ehlers, J. Falco, *Gravitational Lenses* (Springer, 1992)
6. P. Jetzer, Gravitational lensing, in *Modern Cosmology*. IOP Series in High Energy Physics, Cosmology and Gravitation, ed. by S. Bonometto, V. Gorini, U. Moschella (2002), pp. 378–419
7. P. Schneider, C. Kochanek, J. Wambsganss, *Gravitational Lensing: Strong, Weak and Micro* (Springer, 2006)

8. D.G. Yakovlev, P. Haensel, G. Baym, C.J. Pethick, Lev Landau and the conception of neutron stars. Phys. Usp. **56**, 289–295 (2013). arXiv: 1210.0862
9. J.R. Oppenheimer, G.M. Volkoff, Phys. Rev. **55**, 374–381 (1939)
10. N. Straumann, *General Relativity* (Springer, 2013)
11. G. Srinivasan, Astron. Astrophys. Rev. **11**, 67–96 (2002)
12. B. Abbott et al., Phys. Rev. D **96**, 122004 (2017)
13. R. Abbott et al., (2021). arXiv:2111.13106
14. I.I. Shapiro, Phys. Rev. Lett. **13**, 789 (1964)
15. B. Bertotti et al., Nature **425**, 374 (2003)
16. P. Demorest et al., Nature **467**, 1081 (2010)
17. C. Everitt et al., Class. Quantum Gravity **25**, 114002 (2008)
18. C. Everitt et al., Phys. Rev. Lett. **106**, 221101 (2011)
19. W. de Sitter, Mon. Not. Roy. Astron. Soc. **77**, 155 (1916)
20. I.I. Shapiro et al., Phys. Rev. Lett. **61**, 2643 (1988)
21. J. Müller et al., Astrophys. J. **382**, L101 (1991)
22. J. Lense, H. Thirring, Phys. Zeitschr. **19**, 156 (1918)
23. C.W.F. Everitt et al., Phys. Rev. Lett. **106**, 221101 (2011)
24. I. Ciufolini, E.C. Pavlis, Nature **431**, 958 (2004)

Chapter 3
Gravitational Waves and Post-Newtonian Approximation

In this chapter we discuss the gravitational waves, which can be derived using the linearized Einstein field equations. We discuss also sources of gravitational waves and the quadrupole radiation formula, first derived by Einstein. We treat then the post-Newtonian approximation, which is valid for moderately relativistic systems, and is used, for instance, to describe binary pulsar systems.

3.1 Introduction

On September 14, 2015, the two LIGO detectors simultaneously observed a transient gravitational wave (GW) signal, which has been interpreted as due to the merger of two black holes with masses of about 36 $M_\odot$ and 29 $M_\odot$, respectively. This being the first direct detection of GW, which was announced on February 11, 2016 [1] and led to the 2017 Nobel prize in physics for Rainer Weiss, Kip Thorne and Barry Barish for "decisive contributions to the LIGO detector and the observation of gravitational waves".[1] Since then in the data of the first, second and third Advanced LIGO/Virgo observing runs some 90 GW events have been found, most of which are due to the coalescence of two black holes. A few coalescence of two neutron stars and neutron star black holes have been found as well [2, 3].

Moreover, the satellite LISA Pathfinder, with the aim to test the technology needed to build LISA, a GW detector in space, was successfully launched on 3 December 2015. On June 7, 2016, the first results which showed that the performance of LISA Pathfinder was much better than expected and almost already at the level of the LISA requirements were released [4, 5]. Thus the year 2016, 100 years after Einstein's first paper on GW as a consequence of his theory of general relativity, has seen dramatic advancements in the field of GW.

---

[1] The Nobel Prize in Physics 2017, http://www.nobelprize.org/prizes/physics/2017.

© The Author(s), under exclusive license to Springer Nature Switzerland AG 2022
P. Jetzer, *Applications of General Relativity*, UNITEXT for Physics,
https://doi.org/10.1007/978-3-030-95718-6_3

For weak gravitational fields (i.e., $|h_{\mu\nu}| = |g_{\mu\nu} - \eta_{\mu\nu}| \ll 1$) the Einstein field equations read (see 2.139)

$$\Box h_{\mu\nu} = -\frac{16\pi G}{c^4}\left(T_{\mu\nu} - \frac{T}{2}\eta_{\mu\nu}\right). \tag{3.1}$$

In the vacuum ($T_{\mu\nu} = 0$) the equation reduces to

$$\Box h_{\mu\nu} = 0, \tag{3.2}$$

which has plane waves as its simplest solution. The above equation is quite similar to the wave equation in electromagnetism, $\Box A^{\mu} = 0$ with the electromagnetic vector potential A^{μ}. As we will see, the solutions are similar, as well. Note that the wave equation in electromagnetism is exact solutions of Maxwell's equations, whereas the general relativistic wave equation arises from the approximate linearized field equations.

3.1.1 Electromagnetic Waves

Physical fields are invariant under gauge transformations, in particular for the four potential A^{μ} in electrodynamics

$$A^{\mu} \rightarrow A'^{\mu} = A^{\mu} + \partial^{\mu}\chi, \tag{3.3}$$

where χ is a scalar function, so that we can choose $\partial_{\mu} A^{\mu} = 0$ (Lorentz gauge) and get

$$\Box A^{\mu} = \frac{4\pi}{c} j^{\mu}. \tag{3.4}$$

Due to the gauge conditions, only three out of four components of A^{μ} are independent. While leaving the Lorentz gauge unaltered, we still have the freedom to perform an additional gauge transformation satisfying $\Box\chi = 0$. Since in vacuum $j^{\mu} = 0$, this allows us to set $A^0 = 0$. Finally we are left with two degrees of freedom (polarizations). The conditions read then

$$\Box A^{\mu} = 0, \qquad A^0 = 0, \qquad \partial_i A^i = 0. \tag{3.5}$$

This is solved by the ansatz

$$A^{\mu} = e^{\mu}\exp[-ik_{\nu}x^{\nu}] + \text{c.c.}, \tag{3.6}$$

where $k_{\mu}k^{\mu} = 0$ and $e_i k^i = 0$ (polarizations are transverse to the propagation direction).

3.2 Gravitational Waves

Due to the symmetry $h_{\mu\nu} = h_{\nu\mu}$, 10 out of 16 components of $h_{\mu\nu}$ are independent. With a gauge transformation of the form (2.138) we can impose four additional conditions. This leaves us with 6 degrees of freedom that are truly independent. If we consider the vacuum case

$$\Box h_{\mu\nu} = 0, \tag{3.7}$$

in addition to (2.137) we can perform a further transformation of the form

$$h_{\mu\nu} \to h'_{\mu\nu} = h_{\mu\nu} - \partial_\mu \varepsilon_\nu - \partial_\nu \varepsilon_\mu, \tag{3.8}$$

provided that ε_μ satisfies

$$\Box \varepsilon_\mu = 0. \tag{3.9}$$

Such a transformation leaves Eq. (3.7) and the gauge condition (2.138) invariant (this is in complete analogy to electromagnetism, of course). With these four additional conditions we are left with two independent components of $h_{\mu\nu}$. The solution to (3.7) can be written in terms of plane waves

$$h_{\mu\nu} = e_{\mu\nu} \exp[-ik_\kappa x^\kappa] + \text{c.c.}, \tag{3.10}$$

where

$$\eta^{\lambda\nu} k_\lambda k_\nu = k^\nu k_\nu = 0 \quad \Leftrightarrow \quad k_0^2 = \frac{\omega^2}{c^2} = \vec{k}^2 = k^2. \tag{3.11}$$

The amplitude of the wave $e_{\mu\nu}$ is called **polarization tensor**. Inserting (3.10) into the gauge condition (2.138) ($2h^\mu{}_{\nu,\mu} = h^\mu{}_{\mu,\nu}$) leads to

$$2k_\mu \eta^{\mu\rho} e_{\rho\nu} = k_\nu \eta^{\mu\rho} e_{\rho\mu}. \tag{3.12}$$

Clearly $e_{\mu\nu}$ inherits the symmetry of $h_{\mu\nu}$, thus $e_{\mu\nu} = e_{\nu\mu}$. Let us choose a wave traveling along the x^3-axis. This yields the wave solution

$$h_{\mu\nu} = e_{\mu\nu} \exp\left[ik(x^3 - ct)\right], \tag{3.13}$$

where we used Eq. (3.11). The components of the wave vector are then

$$k_1 = k_2 = 0, \qquad k_0 = -k_3 = k = \frac{\omega}{c}. \tag{3.14}$$

In this case the gauge condition (3.12) reads

$$e_{00} + e_{30} = \frac{1}{2}(e_{00} - e_{11} - e_{22} - e_{33}), \tag{3.15}$$

$$e_{01} + e_{31} = 0, \tag{3.16}$$

$$e_{02} + e_{32} = 0, \tag{3.17}$$

$$e_{03} + e_{33} = -\frac{1}{2}(e_{00} - e_{11} - e_{22} - e_{33}). \tag{3.18}$$

With $e_{\mu\nu} = e_{\nu\mu}$ and these four conditions, the polarization tensor is fully determined by six components. All the other components can be expressed in terms of the six independent components

$$e_{00}, \; e_{11}, \; e_{33}, \; e_{12}, \; e_{13} \text{ and } e_{23}. \tag{3.19}$$

The other components are given by

$$e_{01} = -e_{31} = -e_{13}, \quad e_{02} = -e_{32}, \quad e_{22} = -e_{11}, \quad e_{03} = -\frac{1}{2}(e_{00} + e_{33}). \tag{3.20}$$

We can perform yet another transformation (2.132) ($x'^{\mu} = x^{\mu} + \varepsilon^{\mu}$) with functions ε^{μ} satisfying $\Box\varepsilon^{\mu} = 0$. The functions ε^{μ} are solutions of the wave equation, therefore we can write them as

$$\varepsilon^{\mu}(x) = \delta^{\mu} \exp\left[-ik_{\mu}x^{\mu}\right] + \text{c.c.}. \tag{3.21}$$

As noted before, such a transformation with arbitrary δ^{μ} does not violate the gauge condition (2.138). We choose k^{μ} in (3.21) equal to the wave vector of a given gravitational wave. Using (3.21) in (2.138) we obtain a new solution $h'_{\mu\nu}$ in which all the terms have the same exponential dependence of $\exp[-ik_{\mu}x^{\mu}]$. Thus only the amplitudes transform as

$$e'_{11} = e_{11}, \tag{3.22}$$

$$e'_{12} = e_{12}, \tag{3.23}$$

$$e'_{13} = e_{13} - i\delta_1 k, \tag{3.24}$$

$$e'_{23} = e_{23} - i\delta_2 k, \tag{3.25}$$

$$e'_{33} = e_{33} - 2i\delta_3 k, \tag{3.26}$$

$$e'_{00} = e_{00} + 2ik\delta_0. \tag{3.27}$$

We can choose δ_{μ} such that $e'_{00} = e'_{13} = e'_{33} = e'_{23} = 0$. This new solution is equivalent to the old one. From the physical point of view, only polarizations corresponding to e'_{11} and e'_{12} are relevant.

Neglecting primes in our notation from now on, we get for the gravitational wave propagating in x^3-direction, after gauging away all redundancies

$$
h_{\mu\nu} = \begin{pmatrix} 0 & 0 & 0 & 0 \\ 0 & e_{11} & e_{12} & 0 \\ 0 & e_{12} & -e_{11} & 0 \\ 0 & 0 & 0 & 0 \end{pmatrix} \cdot \exp\left[ik(x^3 - ct)\right] + \text{c.c.} \tag{3.28}
$$

3.2.1 Helicity of Gravitational Waves

$\vec{k}$ is the x^3-axis direction. We can now see how (3.28) transforms under a rotation around this axis. Since we are in an almost Minkowskian metric we can realize this transformation as a Lorentz transformation described by the matrix

$$
\Lambda^{\mu}{}_{\nu} = \begin{pmatrix} 1 & 0 & 0 & 0 \\ 0 & \cos\varphi & \sin\varphi & 0 \\ 0 & -\sin\varphi & \cos\varphi & 0 \\ 0 & 0 & 0 & 1 \end{pmatrix}. \tag{3.29}
$$

Therefore the polarization tensor transforms as

$$
e'_{\mu\nu} = \Lambda^{\rho}{}_{\mu}\Lambda^{\sigma}{}_{\nu}e_{\rho\sigma}. \tag{3.30}
$$

This yields

$$
e'_{11} = e_{11}\cos(2\varphi) + e_{12}\sin(2\varphi), \tag{3.31}
$$
$$
e'_{12} = -e_{11}\sin(2\varphi) + e_{12}\cos(2\varphi). \tag{3.32}
$$

If we consider $e_{\pm} \equiv e_{11} \pm ie_{12}$ instead, we thus get

$$
e'_{\pm} = e^{\pm 2i\varphi}e_{\pm}. \tag{3.33}
$$

The vectors $e_{\pm}$ have **helicity** ± 2, whereas the wave solutions in electrodynamics have helicity ± 1. Generalizing from the electromagnetic field, which is quantized using a spin 1 particle, the photon, one can thus expect the quanta of the gravitational field to be spin 2 particles. While there is neither evidence for their existence nor a closed theory of quantum gravity, the hypothetical quanta of the gravitational field are commonly referred to as **gravitons**.

3.2.2 Particles in the Field of a Gravitational Wave

Similarly to electromagnetic waves also gravitational waves exert forces on massive particles. We want to explore how the positions of particles are affected in the field of a gravitational wave. For this analysis consider a plane wave along the x^3-direction

$$h_{\mu\nu}(x^3, t) = \begin{pmatrix} 0 & 0 & 0 & 0 \\ 0 & e_{11} & e_{12} & 0 \\ 0 & e_{12} & -e_{11} & 0 \\ 0 & 0 & 0 & 0 \end{pmatrix} \exp\left(ik(x^3 - ct)\right) + \text{c.c.}. \qquad (3.34)$$

The corresponding metric has the form

$$ds^2 = \left(\eta_{\mu\nu} + h_{\mu\nu}(x^3, t)\right) dx^\mu dx^\nu. \qquad (3.35)$$

The trajectory $x^\sigma(\tau)$ of a particle in the gravitational field satisfies the equation of motion

$$\frac{d^2 x^\sigma}{d\tau^2} = -\Gamma^\sigma{}_{\mu\nu} \frac{dx^\mu}{d\tau} \frac{dx^\nu}{d\tau}, \qquad (3.36)$$

where $\Gamma^\sigma{}_{\mu\nu}$ can be computed with the metric given in Eq. (3.28). We assume that there are no other forces but gravity which act on the particles. Inserting (3.34) into

$$\Gamma^\sigma{}_{\mu\nu} = \frac{1}{2} \eta^{\sigma\lambda} \left(\frac{\partial h_{\nu\lambda}}{\partial x^\mu} + \frac{\partial h_{\mu\lambda}}{\partial x^\nu} - \frac{\partial h_{\mu\nu}}{\partial x^\lambda} \right) + \mathcal{O}\left(h^2\right), \qquad (3.37)$$

it follows

$$\Gamma^i{}_{00} = -\frac{1}{2} \left(\frac{\partial h_{0i}}{\partial x^0} + \frac{\partial h_{0i}}{\partial x^0} - \frac{\partial h_{00}}{\partial x^i} \right) = 0. \qquad (3.38)$$

As initial conditions we choose

$$\dot{x}^i(0) = \frac{dx^i}{d\tau}\bigg|_{\tau=0} = 0. \qquad (3.39)$$

This implies

$$\frac{d^2 x^i}{d\tau}\bigg|_{\tau=0} = -\Gamma^i{}_{\mu\nu} \dot{x}^\mu(0) \dot{x}^\nu(0) \overset{(3.39)}{=} 0. \qquad (3.40)$$

Therefore the acceleration vanishes. This means that the velocities of the particles don't change. The solution of the equations of motion (3.36) thus reads

$$\frac{\mathrm{d}x^i}{\mathrm{d}\tau} = 0 \quad \Rightarrow \quad x^i(\tau) = \text{const.} \tag{3.41}$$

In the chosen coordinates the particles in the field of the gravitational wave can thus be described by constant spatial coordinates. However, this does not mean that the particles are at rest. In fact their distances vary due to the time dependence of the metric tensor $g_{\mu\nu}$ as in Eq. (3.35).

Consider now particles which are arranged on a circle in the x^1-x^2-plane. The particles are initially at rest on a circle $((x^1)^2 + (x^2)^2 = R^2)$. We want to examine the effect of an incident gravitational wave along the x^3-axis. To do so, we write (3.35) in the form

$$ds^2 = c^2\mathrm{d}t^2 - \mathrm{d}l^2 - (\mathrm{d}x^3)^2$$
$$\text{with } \mathrm{d}l^2 = (\delta_{mn} - h_{mn}(t))\,\mathrm{d}x^m\mathrm{d}x^n \quad (m, n = 1, 2). \tag{3.42}$$

In the x^1-x^2-plane we have

$$h_{mn}(t) = h_{mn}(x^3 = 0, t) = \begin{pmatrix} e_{11} & e_{12} \\ e_{12} & -e_{11} \end{pmatrix} \exp(-i\omega t) + \text{c.c.}, \tag{3.43}$$

where $\omega^2 = c^2k^2$. With (3.42) we can compute the physical distance ρ of a particle p from the center of the circle. According to Eq. (3.41) the coordinates x_p^1 and x_p^2 of the particle are constant. We insert in (3.42) the finite values of the coordinates x_p^m of p instead of $\mathrm{d}x^m$ (this is allowed because the metric coefficients do not depend on x^1 and x^2):

$$\rho^2 = (\delta_{mn} - h_{mn}(t))\,x_p^m x_p^n \quad (m, n = 1, 2) \tag{3.44}$$
$$\text{with } x_p^1 = R\cos\varphi, \quad x_p^2 = R\sin\varphi. \tag{3.45}$$

Using Eqs. (3.43)–(3.45) we find the solution

$$\rho^2 = R^2 \begin{cases} 1 - 2h\cos(2\varphi)\cos(\omega t) & \text{if } e_{11} = h, \ e_{12} = 0, \\ 1 - 2h\sin(2\varphi)\cos(\omega t) & \text{if } e_{11} = 0, \ e_{12} = h. \end{cases} \tag{3.46}$$

The $\cos(\omega t)$ term comes from $e^{-i\omega t} + \text{c.c.}$, whereas the $\cos(2\varphi)$ term, for example, comes from $\cos^2\varphi - \sin^2\varphi = \cos(2\varphi)$. The distinction that we made in (3.46) concerns the two possible linear polarization states.

Unlike to the coordinates x^1, x^2 which are constant, the physical variables $x = \rho\cos\varphi$ and $y = \rho\sin\varphi$ describe the distance relative to the center. The physical oscillations lead to a particle configuration which is an ellipse with very tiny eccentricities ($h \ll 1$). From the type of oscillation one can infer the polarization of the incoming wave. The two independent polarization states form an angle of $\frac{\pi}{4}$. Therefore the oscillations correspond to a quadrupole moment of the mass distribution:

The gravitational waves induce a quadrupole oscillation of the mass distribution. Conversely we expect that mass distributions with oscillating quadrupole moment should emit gravitational waves. In order to study this phenomenon further, we have to learn more about the energy and momentum of a gravitational wave.

3.2.3   Energy and Momentum of a Gravitational Wave

We now want to determine the energy-momentum tensor of a gravitational wave, which is a solution of the free field equation up to first order in h ($|h_{\mu\nu}| \ll 1$)

$$R^{(1)}_{\mu\nu} = 0. \tag{3.47}$$

The solution to this equation is the wave solution that we derived before:

$$h_{\mu\nu} = e_{\mu\nu} \exp\left(-ik_\lambda x^\lambda\right) + \text{c.c.}. \tag{3.48}$$

The energy-momentum tensor of a gravitational field is known from Eq. (2.132):

$$t^{\text{grav.}}_{\mu\nu} = \frac{c^4}{8\pi G} \left(R^{(2)}_{\mu\nu} - \frac{\left(g_{\mu\nu} R\right)^{(2)}}{2} \right), \tag{3.49}$$

with $g_{\mu\nu} = \eta_{\mu\nu} + h_{\mu\nu}$. Using $R^{(0)}_{\mu\nu} = R^{(1)}_{\mu\nu} = 0$ and $R = g^{\rho\sigma} R_{\rho\sigma}$ we get

$$
\begin{aligned}
t^{\text{grav.}}_{\mu\nu} &= \frac{c^4}{16\pi G} \left[2R^{(2)}_{\mu\nu} - \eta_{\mu\nu}\eta^{\rho\sigma} R^{(2)}_{\sigma\rho} + \eta_{\mu\nu}h^{\rho\sigma} R^{(1)}_{\sigma\rho} - h_{\mu\nu}\eta^{\rho\sigma} R^{(1)}_{\sigma\rho} \right] \\
&= \frac{c^4}{16\pi G} \left[2R^{(2)}_{\mu\nu} - \eta_{\mu\nu}\eta^{\rho\sigma} R^{(2)}_{\sigma\rho} \right],
\end{aligned}
\tag{3.50}
$$

where we used $g^{\mu\nu} = \eta^{\mu\nu} - h^{\mu\nu}$ which follows from $g^{\mu\nu} g_{\nu\lambda} = \delta^\mu_\lambda$. In order to consider this expression further, we need the Ricci tensor $R^{(2)}_{\mu\nu}$. In order to be able to calculate the Ricci tensor

$$R^{(2)}_{\mu\kappa} = \left(g^{\lambda\nu} R_{\lambda\mu\nu\kappa}\right)^{(2)} = \eta^{\lambda\nu} R^{(2)}_{\lambda\mu\nu\kappa} - h^{\lambda\nu} R^{(1)}_{\lambda\mu\nu\kappa}, \tag{3.51}$$

we need the Riemann tensor which is given by (see Eq. 1.11):

$$
\begin{aligned}
R_{\lambda\mu\nu\kappa} = \frac{1}{2} & \left(\frac{\partial^2 g_{\lambda\nu}}{\partial x^\mu \partial x^\kappa} + \frac{\partial^2 g_{\mu\kappa}}{\partial x^\lambda \partial x^\nu} - \frac{\partial^2 g_{\mu\nu}}{\partial x^\lambda \partial x^\kappa} - \frac{\partial^2 g_{\lambda\kappa}}{\partial x^\mu \partial x^\nu} \right) \\
& + g_{\eta\sigma} \left(\Gamma^\eta_{\nu\lambda} \Gamma^\sigma_{\mu\kappa} - \Gamma^\eta_{\kappa\lambda} \Gamma^\sigma_{\mu\nu} \right),
\end{aligned}
\tag{3.52}
$$

to first and second orders. The first line of this expression (which contains only first order terms in h) gives rise to the second term in (3.51). The second line (which is of second order in h) gives rise to the first term in (3.51). The Christoffel symbols to first order in h are given as

$$\Gamma^{\sigma\,(1)}_{\mu\nu} = \frac{1}{2}\left(\frac{\partial h^{\sigma}{}_{\mu}}{\partial x^{\nu}} + \frac{\partial h^{\sigma}{}_{\nu}}{\partial x^{\mu}} - \frac{\partial h_{\mu\nu}}{\partial x_{\sigma}}\right). \tag{3.53}$$

This yields the first term in (3.51), which is multiplied with $\eta^{\lambda\nu}$. As a result, we get the following expression for the Ricci tensor that we searched for, in order to evaluate Eq. (3.50):

$$
\begin{aligned}
R^{(2)}_{\mu\kappa} = &-\frac{h^{\lambda\nu}}{2}\left[\frac{\partial^2 h_{\lambda\nu}}{\partial x^{\mu}\partial x^{\kappa}} + \frac{\partial^2 h_{\mu\kappa}}{\partial x^{\lambda}\partial x^{\nu}} - \frac{\partial^2 h_{\mu\nu}}{\partial x^{\lambda}\partial x^{\kappa}} - \frac{\partial^2 h_{\lambda\kappa}}{\partial x^{\mu}\partial x^{\nu}}\right] \\
&+ \frac{1}{4}\left[\frac{\partial h^{\nu}{}_{\sigma}}{\partial x^{\nu}} + \frac{\partial h^{\nu}{}_{\sigma}}{\partial x^{\nu}} - \frac{\partial h^{\nu}{}_{\nu}}{\partial x^{\sigma}}\right]\left[\frac{\partial h^{\sigma}{}_{\mu}}{\partial x^{\kappa}} + \frac{\partial h^{\sigma}{}_{\kappa}}{\partial x^{\mu}} - \frac{\partial h_{\mu\kappa}}{\partial x_{\sigma}}\right] \\
&- \frac{1}{4}\left[\frac{\partial h_{\sigma\kappa}}{\partial x^{\lambda}} + \frac{\partial h_{\sigma\lambda}}{\partial x^{\kappa}} - \frac{\partial h_{\lambda\kappa}}{\partial x^{\sigma}}\right]\left[\frac{\partial h^{\sigma}{}_{\mu}}{\partial x_{\lambda}} + \frac{\partial h^{\sigma\lambda}}{\partial x^{\mu}} - \frac{\partial h^{\lambda}{}_{\mu}}{\partial x_{\sigma}}\right]. \tag{3.54}
\end{aligned}
$$

The first term in the second line of this expression vanishes because of the gauge condition (2.138). The remaining terms are quadratic in h and of the form

$$\left[e_{\mu\nu}\exp(-ik_{\lambda}x^{\lambda}) + \text{c.c.}\right]\left[e_{\sigma\kappa}\exp(-ik_{\lambda}x^{\lambda}) + \text{c.c.}\right]$$
$$= e_{\mu\nu}e_{\sigma\kappa}\left[\exp(2ix_{\lambda}k^{\lambda}) + 2 + \exp(-2ix_{\lambda}k^{\lambda})\right]. \tag{3.55}$$

We can see that there are, on the one hand, oscillating terms of the form $\exp(\pm 2ik_{\lambda}x^{\lambda})$ and, on the other hand, there are also terms which do not depend on the coordinates x^{μ} at all. If we average over time, the oscillating terms drop out (their average over time is zero) such that we are left with terms of the form

$$\left\langle\left[e_{\mu\nu}\exp(-ik_{\lambda}x^{\lambda}) + \text{c.c.}\right]\left[e_{\sigma\kappa}\exp(-ik_{\lambda}x^{\lambda}) + \text{c.c.}\right]\right\rangle = 2\Re\left(e^{*}_{\mu\nu}e_{\sigma\kappa}\right), \tag{3.56}$$

where $\langle\cdot\rangle$ denotes time average and $\Re$ is the real part. For plane wave solutions, derivatives correspond to multiplication with k, so

$$\frac{\partial h_{\mu\nu}}{\partial x^{\lambda}} = -ik_{\lambda}h_{\mu\nu}. \tag{3.57}$$

We can now plug all these terms which are quadratic in h into Eq. (3.54), replacing all partial derivatives by factors of k:

$$\langle R^{(2)}_{\mu\kappa} \rangle = \Re\left[(e^{\lambda\nu})^* \left(k_\mu k_\kappa e_{\lambda\nu} + k_\lambda k_\nu e_{\mu\kappa} - k_\lambda k_\kappa e_{\mu\nu} - k_\mu k_\nu e_{\lambda\kappa} \right) \right.$$

$$+ \left(e^\lambda{}_\rho k_\lambda - \frac{1}{2} e_\lambda{}^\lambda k_\rho \right)^* \left(k_\mu e^\rho{}_\kappa + k_\kappa e^\rho{}_\mu - k^\rho e_{\mu\kappa} \right) \tag{3.58}$$

$$\left. - \frac{1}{2} \left(k_\lambda e_{\sigma\kappa} + k_\kappa e_{\sigma\lambda} - k_\sigma e_{\kappa\lambda} \right)^* \cdot \left(k^\lambda e^\sigma{}_\mu + k_\mu e^{\sigma\lambda} - k^\sigma e^\lambda{}_\mu \right) \right].$$

Using the relation: $2\partial_\mu h^\mu{}_\nu = \partial_\nu h^\mu{}_\mu$ [which follows from the gauge condition (2.138)] one can simplify the above expressions. For instance:

$$(e^{\lambda\mu})^* k_\kappa k_\lambda e_{\mu\nu} = \frac{1}{2}(e^\lambda{}_\lambda)^* k^\mu k_\kappa e_{\mu\nu} = \frac{1}{4} k_\kappa k_\nu \left| e^\lambda{}_\lambda \right|^2 . \tag{3.59}$$

Imposing the null condition $k_\mu k^\mu = 0$, we obtain

$$\langle R^{(2)}_{\mu\kappa} \rangle = \frac{1}{2} k_\mu k_\kappa \left[(e^{\lambda\nu})^* e_{\lambda\nu} - \frac{1}{2} |e^\lambda{}_\lambda|^2 \right]. \tag{3.60}$$

Thus the energy-momentum tensor (3.49) or (3.50) of the gravitational wave reads

$$\boxed{t^{\text{grav.}}_{\mu\nu} = \frac{c^4}{16\pi G} k_\mu k_\nu \left[(e^{\lambda\kappa})^* e_{\lambda\kappa} - \frac{1}{2} |e^\lambda{}_\lambda|^2 \right],} \tag{3.61}$$

where we used that

$$\eta_{\mu\nu} \eta^{\rho\sigma} \langle R^{(2)}_{\rho\sigma} \rangle \propto \eta_{\mu\nu} \eta^{\rho\sigma} k_\rho k_\sigma = \eta_{\mu\nu} k^\sigma k_\sigma = 0. \tag{3.62}$$

We can further simplify the energy-momentum tensor by specializing to the case of linearly polarized waves with either $e_{11} = -e_{22} = h, e_{12} = e_{21} = 0$ or $e_{11} = -e_{22} = 0, e_{12} = e_{21} = h$:

$$\boxed{t^{\text{grav.}}_{\mu\nu} = \frac{c^4}{8\pi G} k_\mu k_\nu h^2.} \tag{3.63}$$

Energy in this formula, being proportional to frequency squared is exactly the type of relation that we would intuitively expect. Furthermore it is clear that $t_{\mu\nu} \propto k_\mu k_\nu$ because t_{0i} is the current of momentum which should be proportional to k_i. We see immediately that measuring such energies will be extremely difficult because h^2 is very small. A wave propagating in the x^3-direction has the wave vector $k_\mu = \left(\frac{\omega}{c}, 0, 0, \frac{\omega}{c} \right)$. The energy current density

$$\Phi_{\text{grav.}} = c t^{03}_{\text{grav.}} = \frac{c^5}{8\pi G} k^0 k^3 h^2 \tag{3.64}$$

for such a wave is then given by

$$\Phi_{\text{grav.}} = \frac{\text{energy}}{\text{time} \cdot \text{surface}} = \frac{c^3}{8\pi G}\omega^2 h^2. \tag{3.65}$$

3.2.4 Quadrupole Radiation

Oscillating charge distributions emit electromagnetic waves. In analogy we expect oscillating mass distributions to emit gravitational waves. We quickly repeat the electromagnetic dipole radiation case before turning to the oscillating mass distributions.

In electromagnetism one finds that an oscillating dipole moment

$$\vec{p}(t) = \vec{p}_0 \exp(-i\omega t) + \text{c.c.} \tag{3.66}$$

emits electromagnetic waves whose power P per solid angle is given by

$$\frac{\mathrm{d}P}{\mathrm{d}\Omega} = \frac{\omega^4}{8\pi c^3}|\vec{p}|^2 \sin^2\theta, \tag{3.67}$$

where θ is the angle between $\vec{p}$ and $\vec{k}$, where $\vec{k}$ is the direction of propagation. This is sketched in Fig. 3.1. The total emitted power can be obtained by integrating in θ:

$$P = \frac{\omega^4}{3c^3}|\vec{p}|^2. \tag{3.68}$$

The computation of the emitted gravitational radiation is similar as in electromagnetism, however more involved since the source terms are rank 2 tensors. We will proceed with the following steps:

Fig. 3.1 Sketch of the waves that are emitted by a dipole in electromagnetism

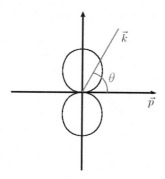

Fig. 3.2 Sketch of the setup
for gravitational wave
emission. The source has
spatial extent r_0. The
observer is at position $\vec{r}$

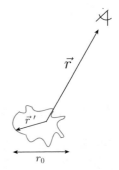

1. Calculate the asymptotic fields emitted by a source $T_{\mu\nu}$.
2. Reduce the result to spatial components.
3. Apply the long wavelength approximation.

The setup is sketched in Fig. 3.2.

In contrast to the electromagnetic case there is no gravitational dipole radiation. The density is given by

$$\rho(\vec{r}, t) = \rho(\vec{r}) \exp(-i\omega t) + \text{c.c.} \quad \Rightarrow \quad \vec{p} = \int d^3r \, \vec{r} \rho(\vec{r}) = M\vec{R}_{\text{c.m.}}, \quad (3.69)$$

where M is the total mass and $\vec{R}_{\text{c.m.}}$ is the center of mass. If we choose the center of mass system as the inertial system then $\vec{p} = 0$. Consequently $\vec{p} = 0$ in all inertial systems. This is not possible in electromagnetism. We shall now assume an oscillatory mass distribution of the form

$$T_{\mu\nu}(\vec{r}, t) = T_{\mu\nu}(\vec{r}) \exp(-i\omega t) + \text{c.c.} \quad \begin{cases} \neq 0 & \text{if } r \leq r_0 \\ = 0 & \text{otherwise.} \end{cases} \quad (3.70)$$

This is only a single Fourier component. Thus a generalization is possible by integrating over ω. According to (2.144) the retarded potentials are given by

$$h_{\mu\nu}(\vec{r}, t) = -\frac{4G}{c^4} \exp(-i\omega t) \int d^3r' \, S_{\mu\nu}(\vec{r}') \frac{\exp(ik|\vec{r} - \vec{r}'|)}{|\vec{r} - \vec{r}'|} + \text{c.c.}, \quad (3.71)$$

where we used

$$-i\omega t_{\text{r}} = -i\omega \left[t - \frac{|\vec{r} - \vec{r}'|}{c} \right] = -i\omega t + ik|\vec{r} - \vec{r}'| \quad (3.72)$$

to obtain the phase factors. Furthermore we have introduced

$$S_{\mu\nu}(\vec{r}) = T_{\mu\nu}(\vec{r}) - \frac{1}{2}\eta_{\mu\nu}T(\vec{r}). \tag{3.73}$$

We now assume $r_0 \ll r$ with $k = \frac{\omega}{c} = \frac{2\pi}{\lambda}$. For large distances we have $|\vec{r}| \gg r_0$ and thus

$$|\vec{r} - \vec{r}'| = r - \frac{\vec{r} \cdot \vec{r}'}{r} + \cdots = r\left[1 + \mathcal{O}\left(\frac{r'}{r}\right)\right] \tag{3.74}$$

and

$$\exp\left[ik|\vec{r} - \vec{r}'|\right] = \exp[ikr]\exp[-i\vec{k}\vec{r}']\left[1 + \mathcal{O}\left(\frac{r'}{r}\right)\right], \tag{3.75}$$

where we defined $\vec{k} = k\frac{\vec{r}}{r} = k\vec{e}_r$. This way we obtain for (3.71)

$$h_{\mu\nu}(\vec{r}, t) = -\frac{4G}{c^4}\frac{1}{r}\exp\left[-ik_\lambda x^\lambda\right]\underbrace{\int d^3r'\, S_{\mu\nu}(\vec{r}')\exp\left[-i\vec{k}\vec{r}'\right]}_{=:S_{\mu\nu}(\vec{k})} + \text{c.c.}, \tag{3.76}$$

where $S_{\mu\nu}(\vec{k})$ is the spatial Fourier transform of $S_{\mu\nu}(\vec{r})$. This yields

$$h_{\mu\nu}(\vec{r}, t) = e_{\mu\nu}(\vec{r}, \omega)\exp\left[-ik_\lambda x^\lambda\right] + \text{c.c.}. \tag{3.77}$$

The amplitudes are defined as

$$e_{\mu\nu}(\vec{r}, \omega) = -\frac{4G}{c^4}\frac{1}{r}S_{\mu\nu}(\vec{k}) = -\frac{4G}{c^4}\frac{1}{r}\left[T_{\mu\nu}(\vec{k}) - \frac{\eta_{\mu\nu}}{2}T(\vec{k})\right] \tag{3.78}$$

and are proportional to $\frac{1}{r}$. They depend on $\frac{\vec{r}}{r} = \vec{e}_r$ and ω via $\vec{k} = k\vec{e}_r$. The energy current passing through a surface element $r^2 d\Omega$ is given by

$$dP = ct_{0i}^{\text{grav.}}df^i = ct_{0i}^{\text{grav.}}\frac{x^i}{r}r^2 d\Omega. \tag{3.79}$$

Plugging in Eq. (3.61) for $t_{\mu\nu}^{\text{grav.}}$ we obtain

$$\frac{dP}{d\Omega} = c\frac{c^4}{16\pi G}\frac{k_0 k_i x^i}{r}r^2\left[(e^{\lambda\nu})^* e_{\lambda\nu} - \frac{1}{2}|e^\lambda{}_\lambda|^2\right]. \tag{3.80}$$

In the derivation of (3.61) we assumed $e_{\mu\nu} = \text{const.}$, whereas here we have $e_{\mu\nu} \propto \frac{1}{r}$. The energy-momentum tensor $t_{\mu\nu}^{\text{grav.}}$ (see Eqs. 3.49, 3.50) contains partial derivatives of the $h_{\mu\nu}$ which lead to additional terms $\propto \frac{1}{r^2}$. With $e_{\mu\nu} = \text{const.}$, the derivatives just lead to factors of $k_\mu \propto \frac{1}{\lambda}$. In the far field and distant observer approximation we have $r \gg \lambda$ and can thus neglect the additional terms $\propto \frac{1}{r^2}$, since $\frac{1}{r} \ll \frac{1}{\lambda}$. Using

$$\frac{k_i x^i}{r} = \frac{\vec{k} \cdot \vec{r}}{r} = k = \frac{\omega}{c} \tag{3.81}$$

in Eq. (3.80) we find that the r^2 factors cancel and we get

$$\frac{dP}{d\Omega} = \frac{G\omega^2}{\pi c^5} \left[T^{\mu\nu}(\vec{k})^* T_{\mu\nu}(\vec{k}) - \frac{1}{2} |T(\vec{k})|^2 \right], \tag{3.82}$$

where $T_{\mu\nu}(\vec{k})$ is the Fourier transform of the source distribution.

We proceed with the second step as outlined in the beginning: We want to reduce our results to spatial components. The source distribution can be written as

$$T^{\mu\nu}(\vec{r}, t) = \frac{1}{(2\pi)^3} \int d^3k \; T^{\mu\nu}(\vec{k}) \exp\left[-ik_\lambda x^\lambda\right] + \text{c.c.}. \tag{3.83}$$

For weak fields the covariant derivative in the energy-momentum conservation simplifies to an ordinary derivative and the continuity equation reads

$$k_\mu T^{\mu\nu}(\vec{k}) = 0. \tag{3.84}$$

In particular we find for $\nu = 0$ and $\nu = i$

$$k_0 T^{00} = -k_j T^{j0} \quad \text{and} \quad k_0 T^{0i} = -k_j T^{ji}. \tag{3.85}$$

We can define a three-dimensional unit vector $\hat{k}_i = \frac{k_i}{k}$ and obtain for (3.85)

$$T^{0i} = T^{i0} = -\hat{k}_j T^{ij}, \tag{3.86}$$

$$T^{00} = \hat{k}_i \hat{k}_j T^{ij}. \tag{3.87}$$

All non-spatial components in (3.82) can thus be eliminated and we compute

$$\begin{aligned}
T^{\mu\nu*} T_{\mu\nu} &= \eta_{\mu\rho} \eta_{\nu\sigma} T^{\mu\nu*} T^{\rho\sigma} \\
&= T^{00*} T^{00} - 2 \sum_i T^{0i*} T^{0i} + \sum_{i,j} T^{ij*} T^{ij} \\
&= \hat{k}_i \hat{k}_j \hat{k}_l \hat{k}_m T^{ij*} T^{lm} - 2\hat{k}_j \hat{k}_m \delta_{il} T^{ij*} T^{lm} + \delta_{il} \delta_{jm} T^{ij*} T^{lm},
\end{aligned} \tag{3.88}$$

$$\begin{aligned}
T^\lambda{}_\lambda &= \eta_{\lambda\rho} T^{\rho\lambda} \\
&= T^{00} - \sum_i T^{ii} \\
&= \hat{k}_i \hat{k}_j T^{ij} - \delta_{ij} T^{ij},
\end{aligned} \tag{3.89}$$

$$|T^\lambda{}_\lambda|^2 = \hat{k}_i \hat{k}_j \hat{k}_l \hat{k}_m T^{ij*} T^{lm} - \delta_{ij} \hat{k}_l \hat{k}_m T^{ij*} T^{lm} - \delta_{lm} \hat{k}_i \hat{k}_j T^{ij*} T^{lm}. \tag{3.90}$$

Inserting these expressions into (3.82) we get

$$\frac{\mathrm{d}P}{\mathrm{d}\Omega} = \frac{G\omega^2}{\pi c^5} \Lambda_{ij,lm} T^{ij}(\vec{k})^* \, T^{lm}(\vec{k}), \tag{3.91}$$

where we introduced

$$\Lambda_{ij,lm}(\theta, \varphi) = \delta_{il}\delta_{jm} - \frac{1}{2}\delta_{ij}\delta_{lm} - 2\delta_{il}\hat{k}_j\hat{k}_m + \frac{1}{2}\delta_{ij}\hat{k}_l\hat{k}_m + \frac{1}{2}\delta_{lm}\hat{k}_i\hat{k}_j + \frac{1}{2}\hat{k}_i\hat{k}_j\hat{k}_l\hat{k}_m. \tag{3.92}$$

Having reduced the formula for the radiated power to spatial components, we now turn to the last step that we outlined in the beginning: We apply the long wavelength approximation, i.e., we assume $\lambda \gg r_0$ which simplifies the energy-momentum tensor as follows:

$$\begin{aligned} T^{ij}(\vec{k}) &= \int \mathrm{d}^3 r \, T^{ij}(\vec{r}) \exp(-i\vec{k}\vec{r}) \\ &= \int \mathrm{d}^3 r \, T^{ij}(\vec{r})(1 - i\vec{k}\vec{r} + \cdots) \\ &\approx \int \mathrm{d}^3 r \, T^{ij}(\vec{r}) \quad =: -\frac{\omega^2}{2} Q^{ij}. \end{aligned} \tag{3.93}$$

The object Q^{ij} will turn out to be a quadrupole tensor. From covariant conservation of energy-momentum, $T^{\mu\nu}{}_{,\nu} = 0$, we get

$$\partial_j T^{ij}(\vec{r}, t) = -\partial_0 T^{i0}(\vec{r}, t) \quad \text{and} \quad \partial_i T^{0i}(\vec{r}, t) = -\partial_0 T^{00}(\vec{r}, t). \tag{3.94}$$

Using Eq. (3.70) we obtain

$$\partial_i\partial_j T^{ij}(\vec{r}, t) = \partial_0^2 \, T^{00}(\vec{r}, t) = -\frac{\omega^2}{c^2} T^{00}(\vec{r}, t) \tag{3.95}$$

$$\Rightarrow \quad \partial_i\partial_j T^{ij}(\vec{r}) = -\frac{\omega^2}{c^2} T^{00}(\vec{r}). \tag{3.96}$$

Since we are in the non-relativistic regime ($\lambda \gg r_0$, $v \ll c$) we have $T^{00} \simeq \rho c^2$. Therefore Eq. (3.93) yields

$$2\int \mathrm{d}^3 r \, T^{ij}(\vec{r}) = 2\int \mathrm{d}^3 r \, x^i x^j \left(\partial_k\partial_l T^{lk}(\vec{r})\right) = -\frac{\omega^2}{c^2} \int \mathrm{d}^3 r \, x^i x^j T^{00}(\vec{r}), \tag{3.97}$$

where we integrated by parts twice in the first step and used Eq. (3.96) in the second step. According to the definition in Eq. (3.93), we find

$$Q^{ij} = \int d^3r \; x^i x^j \rho(\vec{r}) = \frac{1}{c^2} \int d^3r \; x^i x^j T^{00}(\vec{r}), \tag{3.98}$$

which we can obviously interpret as the quadrupole tensor of the mass distribution.[2] Because we are in almost Minkowskian spacetime, we can compute Q^{ij} in three-dimensional Cartesian coordinate. Inserting (3.93) into (3.91) we get

$$\frac{dP}{d\Omega} = \frac{G\omega^6}{4\pi c^5} \Lambda_{ij,lm} Q^{ij*} Q^{lm}. \tag{3.99}$$

We observe that the corresponding formula in electrodynamics looks very similar but it depends on ω^4 rather than ω^6. This is just a reflection of the fact that the electromagnetic radiation is dipole radiation whereas gravitational dipole radiation does not exist (the dipole moment of any mass distribution vanishes in the center of mass system).

Furthermore we note that Q^{ij} are constants (they do not depend on θ or φ). The complete angular dependence is encoded in $\Lambda_{ij,lm}$ in which the vector $\hat{k}$ appears. This is the unit vector which indicates the direction from the mass distribution to the observer and therefore clearly depends on θ and φ:

$$(\hat{k}^i) = (\hat{k}_x, \hat{k}_y, \hat{k}_z) = (\sin\theta \cos\varphi, \sin\theta \sin\varphi, \cos\theta). \tag{3.100}$$

This simplifies calculations, of course, because the quadrupole moment can be calculated once and forever and the specific angular dependence is only to be considered in the form of $\Lambda_{ij,lm}$.

As an example we consider a quadrupole mass distribution in the principal axis system: The only non-vanishing elements are the diagonal, $Q_{ij} = \text{diag}(Q_{11}, Q_{22}, Q_{33})$. In this case the only non-vanishing terms in (3.99) are those with $i = j$ and $l = m$. This means that there appear only even powers of $\hat{k}_x, \hat{k}_y, \hat{k}_z$ in Eq. (3.92). The emitted power has an angular dependence of the form

$$\frac{dP}{d\Omega} \sim a_1 \cos^4\theta + a_2 \cos^2\theta \sin^2\theta + a_3 \sin^4\theta. \tag{3.101}$$

To get the total emitted power, we have to integrate Eq. (3.99) over $d\Omega$:

$$\int d\Omega \; \Lambda_{ij,lm} = \frac{2\pi}{15} \left(11\delta_{il}\delta_{jm} - 4\delta_{ij}\delta_{lm} + \delta_{im}\delta_{jl} \right), \tag{3.102}$$

where we used

---

[2] In the literature one also finds different definitions of this tensor. For example, one can define a traceless version where $x^i x^j$ is replaced by $x^i x^j - \frac{1}{3}r^2\delta^{ij}$.

$$\int d\Omega \, \hat{k}_i \hat{k}_j = \frac{4\pi}{3} \delta_{ij}, \tag{3.103}$$

$$\int d\Omega \, \hat{k}_i \hat{k}_j \hat{k}_l \hat{k}_m = \frac{4\pi}{15} \left(\delta_{ij}\delta_{lm} + \delta_{il}\delta_{jm} + \delta_{im}\delta_{jl} \right). \tag{3.104}$$

Inserting this result into (3.99) we obtain for the total emitted power

$$P = \int d\Omega \frac{dP}{d\Omega} = \frac{2G\omega^6}{5c^5} \left(\sum_{i,j=1}^{3} |Q^{ij}|^2 - \frac{1}{3} \left| \sum_{i=1}^{3} Q^{ii} \right|^2 \right). \tag{3.105}$$

Note that Q_{ij} can be defined traceless such that the second sum in the brackets vanishes. Furthermore one can assume a more general time dependence than just $e^{-i\omega t}$. If we had defined the quadrupole moments in a traceless form,

$$Q^{ij}(t) = \int d^3x \left(x^i x^j - \frac{1}{3} r^2 \delta^{ij} \right) \rho(t, \vec{x}), \tag{3.106}$$

then instead of Eq. (3.105) we would have found

$$P = \frac{G}{5c^5} \langle \dddot{Q}^{kl} \dddot{Q}_{kl} \rangle, \tag{3.107}$$

where $\langle \cdot \rangle$ denotes a time average, for instance over one orbital period.[3] The third time derivatives $\dddot{Q}_{ij}$ in the above equation can be easily evaluated for a plane wave $Q_{ij} \propto \exp[-i\omega t]$ and yield $\dddot{Q}_{ij} \propto \omega^3$.

3.3 Sources of Gravitational Waves

We consider different physical systems whose dynamics leads to the emission of gravitational waves.

3.3.1 Rigid Rotator

As a first example we consider the emission of gravitational waves by a rigid rotating body. Consider a coordinate system KS' with coordinates x'_m in which the body is fixed. In KS' the mass density $\rho'(\vec{r}')$ is time independent. We choose KS' such that the quadrupole tensor Θ'_{ij} is diagonal:

---

[3] In the literature, one finds also the formula $P = \frac{G}{45c^5} \langle \dddot{Q}^{kl} \dddot{Q}_{kl} \rangle$. The different prefactor arises if one defines the quadrupole tensor with an additional factor of 3 in the second term of Eq. (3.106).

$$\Theta'_{ij} = \int d^3r' \; x'_i x'_j \rho'(\vec{r}') = \begin{pmatrix} I_1 & 0 & \\ 0 & I_2 & 0 \\ 0 & 0 & I_3 \end{pmatrix}. \tag{3.108}$$

We assume that the body rotates with angular velocity Ω around the x'_3-axis. The orthogonal transformation to an inertial frame IS with coordinates x_n can be written as

$$x_n = \alpha^m{}_n(t) x'_m \quad \text{with } \alpha^m{}_n(t) = \begin{pmatrix} \cos \Omega t & -\sin \Omega t & 0 \\ \sin \Omega t & \cos \Omega t & 0 \\ 0 & 0 & 1 \end{pmatrix}. \tag{3.109}$$

The tensor Θ_{ij} in IS reads

$$\begin{aligned} \Theta_{ij}(t) &= \int d^3r \; x_i x_j \rho(\vec{r}, t) \\ &= \int d^3r' \; \left(\alpha^n{}_i x'_n \right) \left(\alpha^m{}_j x'_m \right) \rho'(\vec{r}') \\ &= (\alpha(t) \Theta' \alpha^{\mathsf{T}}(t))_{ij}, \end{aligned} \tag{3.110}$$

where we used $d^3r = d^3r'$ and $\rho'(\vec{r}') = \rho(\vec{r})$ since the density transforms as a scalar quantity. With Eqs. (3.108) and (3.109) we can compute (3.110):

$$\begin{aligned} \Theta_{11}(t) &= \frac{I_1 + I_2}{2} + \frac{I_1 - I_2}{2} \cos(2\Omega t) \\ \Theta_{12}(t) &= \frac{I_1 - I_2}{2} \sin(2\Omega t) \\ \Theta_{22}(t) &= \frac{I_1 + I_2}{2} - \frac{I_1 - I_2}{2} \cos(2\Omega t) \\ \Theta_{33}(t) &= I_3, \qquad \Theta_{13}(t) = \Theta_{31}(t) = 0. \end{aligned} \tag{3.111}$$

This is of the form

$$\Theta_{ij} = \text{const.} + \left[Q_{ij} \exp(-2i\Omega t) + \text{c.c.} \right] \tag{3.112}$$

$$\text{with } Q_{ij} = \frac{I_1 - I_2}{4} \begin{pmatrix} 1 & i & 0 \\ i & -1 & 0 \\ 0 & 0 & 0 \end{pmatrix}. \tag{3.113}$$

Comparing Eqs. (3.110) and (3.112) we see that a rotating rigid body can be interpreted as a mass distribution whose rotational frequency Ω leads to gravitational waves of frequency $\omega = 2\Omega$. We introduce the moment of inertia I with respect to the rotation axis and the **ellipticity** of the body ε,

$$I = I_1 + I_2, \qquad \varepsilon = \frac{I_1 - I_2}{I_1 + I_2}. \tag{3.114}$$

We can then write

$$P = \frac{32G\Omega^6}{5c^5}\varepsilon^2 I^2,$$

(3.115)

which is what one expects for the quadrupole radiation $(Q \sim \epsilon I)$.

3.3.2 Example: The Binary Star System

As an example we consider a binary star system (masses M_1 and M_2) orbiting around each other on a circular orbit. Assuming a circle with constant radius r, we can consider the system as a rigid rotator:

$$I \simeq I_1 = \frac{M_1 M_2}{M_1 + M_2} r^2, \qquad I_2 \simeq 0, \qquad \varepsilon \simeq 1.$$

(3.116)

The circular orbit is characterized by

$$\underbrace{\frac{M_1 M_2}{M_1 + M_2}}_{=\mu} \Omega^2 r = \frac{G M_1 M_2}{r^2} \qquad \Rightarrow \qquad \Omega^2 = \frac{GM}{r^3},$$

(3.117)

where μ is the reduced mass and $M = M_1 + M_2$. Inserting this into Eq. (3.115) we find

$$P = \frac{32 G^4 M_1^2 M_2^2 (M_1 + M_2)}{5c^5 r^5} = \frac{32}{5} \frac{G^4}{c^5 r^5} M^3 \mu^2.$$

(3.118)

It is convenient to express the emitted power in terms of the Planck luminosity

$$L_P = \frac{c^5}{G} = 3.63 \times 10^{59}\ \text{erg s}^{-1}$$

$$= 2.03 \times 10^5 M_\odot c^2\ \text{s}^{-1},$$

(3.119)

which yields

$$P = \frac{32}{5} \left(\frac{GM}{c^2 r}\right)^5 \frac{\mu^2}{M^2} L_P.$$

(3.120)

For example, coalescing neutron stars in the final stage have $r \approx R_S$ and thus $\frac{GM}{c^2 r} \sim \mathcal{O}(1)$ and $P \approx L_P$. In general, for order of magnitude estimates one can use

$$P = L_P \left(\frac{GM}{c^2 R} \right)^5 \tag{3.121}$$

for systems of typical scale R.

Due to the emission of gravitational waves the system loses energy and thus its distance R shrinks, until the two bodies coalesce after a time $t_{\text{spir.}}$ (which corresponds to the inspiral time). In the Kepler problem the total energy is

$$E = -\frac{GM_1 M_2}{2r}. \tag{3.122}$$

During the inspiral process, the system loses potential energy which is emitted in the form of gravitational waves. Thus $dE = -Pdt$ and

$$P = -\frac{dE}{dt} = -\frac{GM_1 M_2}{2r^2} \frac{dr}{dt} = \frac{32}{5} \frac{G^4}{c^5} \frac{M_1^2 M_2^2 (M_1 + M_2)}{r^5}. \tag{3.123}$$

Upon substituting $x(t) = [r(t)/r(0)]^4$ we can rewrite the last equality in (3.123) as

$$\frac{dx}{dt} = -\frac{256G^3}{5c^5} \frac{M_1 M_2 (M_1 + M_2)}{r^4(0)} \equiv -\frac{1}{t_{\text{spir.}}}. \tag{3.124}$$

This is solved by $x = 1 - t/t_{\text{spir.}}$ such that we find

$$\boxed{r(t) = r(0) \left(1 - \frac{t}{t_{\text{spir.}}} \right)^{1/4}.} \tag{3.125}$$

Next, we want to calculate the **strain amplitude**[4] of such a system. We evaluate the expression (3.78) for $e_{\mu\nu}$ in analogy to what we did with Eq. (3.79). We can express $T_{\mu\nu}$ in terms of its spatial components ($k_0 T^{0i} = k_j T^{ij}$ and $k_0 T^{00} = k_i T^{0i}$) which in turn can be expressed in terms of Q_{ij}:

$$T_{ij}(\vec{k}) = -\frac{\omega^2}{2} Q_{ij}. \tag{3.126}$$

With the definitions in (3.113) and (3.115) we obtain

$$e_{11} = \frac{1}{2} \frac{GI\varepsilon}{c^4} (2\Omega)^2 \frac{1}{D}, \tag{3.127}$$

where D is the distance between source and observer.

---

[4] The strain amplitude is fractional change in the distance between two measurement points due to the deformation of spacetime by a passing gravitational wave.

If there are two polarizations and $e_{11} = e_{12}$ then we have for the dimensionless strain amplitude

$$h = \sqrt{e_{11}^2 + e_{12}^2} = 2\sqrt{2}\frac{GI\varepsilon}{c^4}\Omega^2\frac{1}{D}. \tag{3.128}$$

For a binary system characterized by masses $\tilde{M} = M_1 = M_2$ on a circular orbit of radius r, this yields

$$r^2\Omega^2 = \frac{G\tilde{M}}{r}, \quad I = \frac{\tilde{M}r^2}{2} \quad \Rightarrow \quad h \sim \frac{R_S^2}{Dr}, \tag{3.129}$$

where $R_S = 2G\tilde{M}/c^2$ is the Schwarzschild radius of the system.

For two neutron stars with $\tilde{M} = 1.4M_\odot$, $r = 100$ km (i.e., $T \approx 10^{-2}$ s, $\Omega \approx 6 \times 10^2$ Hz) at a distance $D = 30000$ ly ≈ 10 kpc we get

$$h \approx 10^{-18} \quad \text{and} \quad P \approx 10^{52} \text{ erg s}^{-1}. \tag{3.130}$$

This strain is the relative amplitude of the oscillation of a ruler's length when the gravitational wave passes through Earth.

3.4 Binaries on Elliptic Orbits

In the previous section we computed the emission of gravitational waves assuming circular orbits for two coalescing compact objects. Here, we consider elliptical orbits (for a more complete discussion, see, e.g., [6]). Let m_1, m_2 be the masses of the compact bodies, such as black holes or neutron stars, at positions $\vec{r}_1$ and $\vec{r}_2$. In the center of mass frame (CM), the problem reduces to a one-body problem for a particle of mass equal to the reduced mass $\mu = \frac{m_1 m_2}{m_1 + m_2}$ and acceleration $\ddot{\vec{r}} = -(G\frac{m}{r^2})\hat{\vec{r}}$, where $m = m_1 + m_2$ and $\vec{r} = \vec{r}_2 - \vec{r}_1$.

3.4.1 Elliptic Keplerian Orbits

There are two integrals of motions, the angular momentum $\vec{L}$ and the energy E. Angular momentum conservation implies that the orbit lies on a plane. We introduce polar coordinates (r, ϕ) in the orbital plane with origin at the CM to describe the orbits. The angular momentum and the energy are given by

$$L = |\vec{L}| = \mu r^2\dot{\phi} \tag{3.131}$$

$$E = \frac{1}{2}\mu(\dot{r}^2 + r^2\dot{\phi}^2) - \frac{G\mu m}{r} = \frac{1}{2}\mu\dot{r}^2 + \frac{L^2}{2\mu r^2} - \frac{G\mu m}{r}. \tag{3.132}$$

From Eq. (3.132) we get $\dot{r}$ as a function of r and from Eq. (3.131) we get $\dot{\phi}$, combining the two expressions we find $\frac{dr}{d\phi}$ as a function of r, and integrating the so obtained expression, we get the equation for the orbit

$$\frac{1}{r} = \frac{1}{R}(1 + e\cos\phi).$$

(3.133)

The *length-scale* R and the *eccentricity* e are constants of motion, related to the energy E of the system[5] and to the orbital angular momentum L by

$$R = \frac{L^2}{Gm\mu^2},$$

(3.134)

and

$$e^2 = 1 + \frac{2EL^2}{G^2 m^2 \mu^3},$$

(3.135)

with $0 \le e < 1$. For $e = 0$ the ellipse becomes a circle, while for $e = 1$ the orbit is a parabola. The two semi-axes of the ellipse are given by

$$a = \frac{R}{1 - e^2},$$

(3.136)

$$b = \frac{R}{\sqrt{(1 - e^2)}}.$$

(3.137)

Inserting the expression for e, as given in Eq. (3.135), we obtain

$$a = \frac{Gm\mu}{2|E|}.$$

(3.138)

Note that a is independent of L, such that orbits with the same energy have the same value of the semi-major axis a. We can rewrite Eq. (3.133) as

$$r = \frac{a(1 - e^2)}{1 + e\cos\phi}.$$

(3.139)

Using Eqs. (3.131) and (3.134), we get

$$\dot{\phi} = \frac{\sqrt{Gm R}}{r^2}.$$

(3.140)

Integrating Eqs. (3.131) and (3.132) leads to $r(t)$ and $\phi(t)$. The parametric solutions are

---

[5] For bound systems, $E < 0$.

$$r = a(1 - e \cos u), \tag{3.141}$$

$$\cos \phi = \frac{\cos u - e}{1 - e \cos u}, \tag{3.142}$$

where u is the *eccentric anomaly*, which is related to t by the *Kepler equation*

$$\beta \equiv u - e \sin u = \omega_0 t, \tag{3.143}$$

where $\omega_0^2 = \frac{Gm}{a^3}$. The origin of time is such that at $t = 0$ we have $\phi = 0$, i.e., the periastron. Using trigonometric identities, Eq. (3.142) can be brought to the form

$$\tan \frac{\phi}{2} = \left(\frac{1 + e}{1 - e} \right)^{\frac{1}{2}} \tan \frac{u}{2}, \tag{3.144}$$

or

$$\phi = A_e(u) = 2 \arctan \left(\left(\frac{1 + e}{1 - e} \right)^{\frac{1}{2}} \tan \frac{u}{2} \right), \tag{3.145}$$

where $A_e(u)$ is called *true anomaly* . Note that in Eq. (3.143) if $t \to t + \frac{2\pi}{\omega_0}$, then we have $\beta \to 2\pi$ and $u \to u + 2\pi$, thus the coordinates r, ϕ are periodic functions of t, with periodicity

$$T = \frac{2\pi}{\omega_0}. \tag{3.146}$$

u has values between $-\pi, \pi$, ϕ is also defined between $-\pi$ and π, and for $e = 0$, $\phi = u$. In Cartesian coordinates, centered on the focus of the ellipse, the orbit is described by

$$x = r \cos \phi = a(\cos u(t) - e), \tag{3.147}$$
$$y = r \sin \phi = b \sin u(t). \tag{3.148}$$

The focus of the ellipse corresponds to the center of mass (Fig. 3.3).

3.4.2 Radiated Power

We compute here the total radiated power in gravitational waves, integrating over all frequencies and the solid angle. We choose a reference frame such that the orbit lies in the (x, y)-plane, this way the *second mass moment* is given by the 2×2 matrix

$$M_{ab} = \mu r^2 \begin{bmatrix} \cos^2 \phi & \sin \phi \cos \phi \\ \sin \phi \cos \phi & \sin^2 \phi \end{bmatrix}_{ab}, \tag{3.149}$$

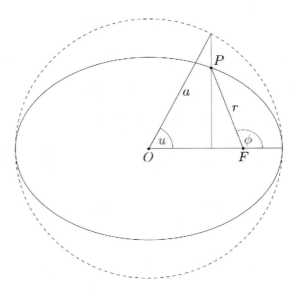

Fig. 3.3 Geometry of the elliptical Kepler orbit. The eccentric anomaly u takes the role of the angle as measured from the center of the ellipse

where $a, b = 1, 2$ denote the indices in the (x, y)-plane. We need the third time derivative of M_{ab} to compute the total power emitted in the quadrupole approximation

$$P = \frac{G}{5c^5} \langle \dddot{M}_{ij} \dddot{M}_{ij} - \frac{1}{3} (\dddot{M}_{kk})^2 \rangle. \tag{3.150}$$

In Eq. (3.149) the time dependence is in $r(t)$ and $\phi(t)$. To compute these derivatives, the simplest way is to write M_{ab} as a function of ϕ only, eliminating r with Eq. (3.139), one gets,

$$M_{11} = \mu r^2 \cos \phi = \mu a^2 (1 - e^2)^2 \frac{\cos \phi}{(1 + e \cos \phi)^2}, \tag{3.151}$$

and

$$\dot{\phi} = \frac{(GmR)^{\frac{1}{2}}}{r^2} = \left(\frac{Gm}{a^3}\right)^{\frac{1}{2}} (1 - e^2)^{-\frac{3}{2}} (1 + e \cos \phi)^2. \tag{3.152}$$

Then we get

$$\dddot{M}_{11} = M_0 (1 + e \cos \phi)^2 (2 \sin(2\phi) + 3e \sin \phi \cos^2 \phi), \tag{3.153}$$

$$\dddot{M}_{22} = M_0 (1 + e \cos \phi)^2 (-2 \sin(2\phi) - e \sin \phi (1 + 3 \cos^2 \phi)), \tag{3.154}$$

$$\dddot{M}_{12} = M_0 (1 + e \cos \phi)^2 (-2 \cos(2\phi) + e \cos \phi (1 - 3 \cos^2 \phi)), \tag{3.155}$$

where

$$M_0^2 = \frac{4G^3 \mu^2 m^3}{a^5 (1 - e^2)^5}. \tag{3.156}$$

Inserting this into the formula for the quadrupole radiation, we get P as a function of the position angle ϕ along the orbit,

$$P(\phi) = \frac{G}{5c^2} \left(\dddot{M}_{11}^2 + \dddot{M}_{22}^2 + 2\dddot{M}_{12}^2 - \frac{1}{3}(\dddot{M}_{11} + \dddot{M}_{22})^2 \right) \tag{3.157}$$

$$= \frac{8G^4}{15c^5} \frac{\mu^2 m^3}{a^5 (1 - e^2)^5} (1 + e \cos \phi)^4 (12(1 + \cos \phi)^2 + e^2 \sin^2 \phi). \tag{3.158}$$

A particle in a Keplerian elliptic orbit emits gravitational waves at frequencies which are integer multiple of the frequency ω_0, defined in Eq. (3.143). The period of the gravitational waves is a fraction of the orbital period T as given in Eq. (3.146). Thus a well-defined quantity is the average of $P(\phi(t))$ over one orbital period T:

$$P = \frac{1}{T} \int_0^T dt\, P(\phi) = \frac{\omega_0}{2\pi} \int_0^{2\pi} \frac{d\phi}{\dot{\phi}} P(\phi(t)) = (1 - e^2)^{\frac{3}{2}} \int_0^{2\pi} \frac{d\phi}{2\phi} \frac{P(\phi)}{(1 + e \cos \phi)^2} =$$

$$= \frac{8G^4 \mu^2 m^3}{15c^5 a^5} (1 - e^2)^{-\frac{7}{2}} \int_0^{2\pi} \frac{d\phi}{2\pi} (12(1 + e \cos \phi)^4 + e^2 (1 + e \cos \phi)^2 \sin^2 \phi).$$

$$\tag{3.159}$$

Evaluating the integral we find [7]

$$P = \frac{32G^4 \mu^2 m^3}{5c^5 a^5} f(e), \tag{3.160}$$

where

$$f(e) = \frac{1}{(1 - e^2)^{\frac{7}{2}}} \left(1 + \frac{73}{24} e^2 + \frac{37}{96} e^4 \right). \tag{3.161}$$

We can eliminate m and rewrite the above result in terms of ω_0

$$P = \frac{32}{5} \frac{G\mu^2}{c^5} a^4 \omega_0^6 f(e). \tag{3.162}$$

Note that for $e = 0$ we have $f(e) = 1$. For the Hulse–Taylor binary pulsar one finds $e = 0.617$ and $f(e) \simeq 11.8$. As a result, the radiated power is an order of magnitude larger than the one emitted if the pulsars would be on a circular orbit with radius a. The fact that the power loss is higher for an elliptic orbit leads to more and more circular orbit for the binary system.

The orbital period T is related to the orbital energy E by $T \sim (-E)^{-\frac{3}{2}}$ and therefore

$$\frac{\dot{T}}{T} = -\frac{3}{2}\frac{\dot{E}}{E}.$$

(3.163)

Using Eq. (3.160), together with Eq. (3.138), to express the energy loss $\dot{E} = P$ in terms of a, we can rewrite Eq. (3.163) as

$$\frac{\dot{T}}{T} = -\frac{96}{5}\frac{G^3\mu m^2}{c^5 a^4} f(e),$$

(3.164)

where averaging over one orbital period has been done. Finally writing a in terms of T, we find

$$\frac{\dot{T}}{T} = -\frac{96}{5}\frac{G^{\frac{5}{3}}\mu m^{\frac{2}{3}}}{c^5}\left(\frac{T}{2\pi}\right)^{-\frac{8}{3}} f(e),$$

(3.165)

which describes the orbital period decay of the binary pulsar system, which lead to the first direct experimental evidence for gravitational radiation emission in the Hulse–Taylor binary pulsar system.

3.4.3 Parabolic Orbits

The limit $e \to 1^-$ of the radiated power given in Eq. (3.160) diverges, since for $e \to 1^-$, keeping a fixed, we get $R \to 0$ (R is related to the minimal scattering distance) and $b = a\sqrt{1 - e^2} \to 0$. Clearly, in this limit the point-like masses approximation is no longer valid and one has to take into account their finite sizes. Consider Eq. (3.133) for the parabolic motion $e = 1$, then

$$r = \frac{R}{1 + \cos\phi}.$$

(3.166)

For $\phi \to -\pi$ the particle is at $r \to \infty$, by increasing ϕ, the value of r decreases down to $r = \frac{R}{2}$ (reached for $\phi = 0$), and then increases again until $r \to +\infty$ as $\phi \to \pi$. In this limit, eliminating a in favor of R using Eq. (3.136), we obtain from Eq. (3.159) the radiated power along the trajectory,

$$\begin{aligned}
P(\phi) &= \frac{8G^4\mu^2 m^3}{15c^5 R^5}(1 + \cos\phi)^4(12(1 + \cos\phi)^2 + \sin^2\phi) \\
&= \frac{16G^4\mu^2 m^3}{15c^5}\frac{1}{r^5}\left(1 + \frac{11R}{2r}\right).
\end{aligned}$$

(3.167)

$P(\phi)$ goes to zero quite fast for $r \to \infty$ (as $\phi \to \pm\pi$) as r^{-5} or r^{-6}, which is expected, since the acceleration decreases as $\frac{1}{r^2}$. Therefore, the total radiated energy in gravitational waves is finite and given by

$$E_{\text{rad}} = \int\limits_{-\infty}^{+\infty} dt\, P(\phi(t)) = \int\limits_{-\pi}^{+\pi} \frac{d\phi}{\dot{\phi}} P(\phi) = \frac{85\pi}{48} \frac{G\mu^2}{R} \left(\frac{v_0}{c}\right)^5, \tag{3.168}$$

where $v_0 = \sqrt{Gm/R}$ is the velocity at $\phi = 0$, which corresponds to $r = \frac{R}{2}$, i.e., the maximum velocity attained along the trajectory. Most of the radiation is emitted for $-\frac{\pi}{2} < \phi < \frac{\pi}{2}$, when the object is near the periastron. Hyperbolic orbits can be treated as well, see, e.g., [8–10].

3.4.4 Frequency Spectrum of Gravitational Waves

We are interested in computing the frequency spectrum of the radiated power, namely $\frac{dP}{d\omega}$ for a Keplerian elliptic orbit. The trajectory is described by Eq. (3.147) and Eq. (3.148) as a function of time. Since it is not a simple harmonic motion, we rewrite the trajectory in terms of Fourier series. Indeed, $x(t)$ and $y(t)$ are periodic functions of the variable $\beta = \omega_0 t$, defined in Eq. (3.143) with period 2π, and thus β can be taken in the range $-\pi \leq \beta \leq \pi$. We define the Fourier series as follows

$$x(\beta) = \sum_{n=-\infty}^{+\infty} \hat{x}_n e^{-in\beta}, \tag{3.169}$$

$$y(\beta) = \sum_{n=-\infty}^{+\infty} \hat{y}_n e^{-in\beta}, v \tag{3.170}$$

with $\hat{x}_n = \hat{x}_{-n}^*$ and $\hat{y}_n = \hat{y}_{-n}^*$, where $*$ indicate the complex conjugate, since $x(\beta)$ and $y(\beta)$ are real functions. We choose the origin of time such that at $t = 0$, i.e., $\beta = 0$, we are at the point $x = a(1-e)$, $y = 0$. Note that the following even and odd properties of the functions $x(-\beta) = x(\beta)$, $y(-\beta) = -y(\beta)$ hold. In the $x(\beta)$ expansion only $\cos(n\beta)$ contribute, whereas in $y(\beta)$ only $\sin(n\beta)$. We rewrite Eqs. (3.169) and (3.170) as

$$x(\beta) = \sum_{n=0}^{\infty} a_n \cos(n\beta), \tag{3.171}$$

$$y(\beta) = \sum_{n=1}^{\infty} b_n \sin(n\beta), \tag{3.172}$$

where, for $n \geq 1$, $a_n = 2\hat{x}_n$ and $b_n = -2i\hat{y}_n$, whereas $a_0 = \hat{x}_0$; with $\beta = \omega_0 t$ and $\omega_n = n\omega_0$ we have

$$x(t) = \sum_{n=0}^{\infty} a_n \cos(\omega_n t), \tag{3.173}$$

$$y(t) = \sum_{n=1}^{\infty} b_n \sin(\omega_n t),$$

(3.174)

where $\omega_0 = \sqrt{(Gm/a^3)}$ is the *fundamental frequency* and ω_n are its higher harmonics. The Fourier coefficients are obtained from

$$a_n = \frac{1}{\pi} \int_{-\pi}^{\pi} d\beta \, x(\beta) \cos(n\beta) \quad (n = 0, 1, \ldots),$$

(3.175)

$$b_n = \frac{1}{\pi} \int_{-\pi}^{\pi} d\beta \, y(\beta) \sin(n\beta), \quad (n = 1, 2, \ldots),$$

(3.176)

where $x(\beta)$, $y(\beta)$ are given by Eqs. (3.147), (3.148) and β by Eq. (3.143) all in terms of $u(t)$. As a result one finds for $n \neq 0$:

$$a_n = \frac{a}{n}(J_{n-1}(ne) - J_{n+1}(ne)),$$

(3.177)

$$b_n = \frac{b}{n}(J_{n-1}(ne) + J_{n+1}(ne)),$$

(3.178)

where J_n is the *Bessel function*, and $a_0 = -3ae$. To compute the spectrum of gravitational waves we actually need the Fourier decomposition of the second mass moment and therefore of $x^2(t)$, $y^2(t)$ and $x(t)y(t)$. One finds

$$x^2(t) = \sum_{n=0}^{\infty} A_n \cos(\omega_n t),$$

(3.179)

$$y^2(t) = \sum_{n=0}^{\infty} B_n \cos(\omega_n t),$$

(3.180)

$$x(t)y(t) = \sum_{n=1}^{\infty} C_n \sin(\omega_n t),$$

(3.181)

with the coefficients given by

$$A_n = \frac{a^2}{n}(J_{n-2}(ne) - J_{n+2}(ne) - 2e(J_{n-1}(ne) - J_{n+1}(ne))),$$

(3.182)

$$B_n = \frac{b^2}{n}(J_{n+2}(ne) - J_{n-2}(ne)),$$

(3.183)

$$C_n = \frac{ab}{n}(J_{n+2}(ne) + J_{n-2}(ne) - e(J_{n+1}(ne) + J_{n-1}(ne))).$$

(3.184)

The Fourier decomposition of the second mass moment is as follows (μ reduced mass)

$$M_{ab}(t) = \mu \sum_{n=0}^{\infty} \begin{bmatrix} A_n \cos(\omega_n t) & C_n \sin(\omega_n t) \\ C_n \sin(\omega_n t) & B_n \cos(\omega_n t) \end{bmatrix}_{ab},$$

$$\equiv \sum_{n=0}^{\infty} M_{ab}^{(n)}(t). \tag{3.185}$$

For computing the radiated power we need the temporal averages such as $\langle \sin(\omega_n t) \cos(\omega_m t) \rangle$, which are non-vanishing only if $n = m$. Therefore, there are no interference terms between the different harmonics, and we can write

$$P = \sum_{n=1}^{\infty} P_n, \tag{3.186}$$

where P_n is the power radiated in the n-th harmonics. P_n are computed using the quadrupole formula, written in the form given in Eq. (3.150)

$$P_n = \frac{2G}{15c^5} \langle \dddot{M}_{11}^{(n)2} + \dddot{M}_{22}^{(n)2} + 3\dddot{M}_{12}^{(n)2} - \dddot{M}_{11}^{(n)} \dddot{M}_{22}^{(n)} \rangle, \tag{3.187}$$

for an orbit in (x, y)-plane, with

$$\dddot{M}_{ab}^{(n)} = \mu \omega_n^3 \begin{bmatrix} A \sin(\omega_n t) & -C_n \cos(\omega_n t) \\ -C_n \cos(\omega_n t) & B_n \sin(\omega_n t) \end{bmatrix}_{ab}. \tag{3.188}$$

We finally get

$$P_n = \frac{G\mu^2 \omega_0^6}{15c^5} n^6 (A_n^2 + B_n^2 + 3C_n^2 - A_n B_n), \tag{3.189}$$

using $\langle \sin^2(\omega_n t) \rangle = \langle \cos^2(\omega_n t) \rangle = \frac{1}{2}$. With $\omega_0^2 = \frac{Gm}{a^3}$ we can rewrite the previous result as

$$P_n = \frac{32}{5} \frac{G\mu^2}{c^5} a^4 \omega_0^6 g(n, e), \tag{3.190}$$

with

$$g(n, e) = \frac{n^6}{96a^4} (A_n^2(e) + B_n^2(e) + 3C_n^2(e) - A_n(e)B_n(e)), \tag{3.191}$$

were the eccentricity is given by $\frac{b^2}{a^2} = (1 - e^2)$. For the limit $e \to 0$, we get back the result for circular motion. For $0 < e < 1$ all harmonics contribute and we have radiation at all frequencies $\omega_n = n\omega_0$, for $n \geq 1$.

3.4.5 Evolution of the Orbit Due to Gravitational Waves Emission

A binary system in a Keplerian orbit looses energy and angular momentum. We assume the bodies as being point-like, without intrinsic spin. Due to energy and angular momentum losses the semi-major axis and the ellipticity change, which we discuss in the following. The energy radiated in quadrupolar approximation has already been computed, see Eqs. (3.160) and (3.162). As next, we compute the angular momentum radiated within quadrupole approximation. It can be shown that (summation over same indices applies)

$$\frac{dL^i}{dt} = -\frac{2G}{5c^5} \varepsilon^{ikl} \langle \ddot{M}_{ka} \dddot{M}_{la} \rangle, \tag{3.192}$$

where L^i ($i = x, y, z$) is the orbital angular momentum of the binary system. The coordinate system is such that the orbit is in the (x, y)-plane. M_{ab} is given by Eq. (3.149) and $L_z = L$ is orthogonal to the orbital plane. Remember that $\langle \ldots \rangle = \frac{1}{T} \int_0^T dt \ldots$ means averaging over the time interval. One can thus integrate Eq. (3.192) by parts and get the following relation

$$\frac{dL}{dt} = -\frac{2G}{5c^5} \langle \ddot{M}_{1a} \dddot{M}_{2a} - \ddot{M}_{2a} \dddot{M}_{1a} \rangle = \frac{4G}{5c^5} \langle \ddot{M}_{12} (\dddot{M}_{11} - \dddot{M}_{22}) \rangle. \tag{3.193}$$

The third derivatives have been computed in Eqs. (3.153)–(3.155). Moreover we need

$$\ddot{M}_{12} = \frac{G\mu m}{a(1 - e^2)} \sin \phi (-4(1 + e \cos \phi)^2 \cos \phi + 2e(3 \cos^2 \phi - 1 + 2e \cos^3 \phi)). \tag{3.194}$$

For a periodic motion averaging over several periods of the wave leads to the same as averaging over one orbital period T. We thus rewrite the temporal average over one period as an integration over ϕ

$$\int_0^T \frac{dt}{T} \ldots = (1 - e^2)^{\frac{3}{2}} \int_0^{2\pi} \frac{d\phi}{2\pi} (1 + e \cos \phi)^{-2} \ldots. \tag{3.195}$$

Finally, we get the averaged angular momentum loss over one period

$$
\frac{dL}{dt} = \frac{8}{5} \frac{G^{\frac{7}{2}} \mu^2 m^{\frac{5}{2}}}{c^5 a^{\frac{7}{2}}} \frac{1}{(1-e^2)^2} \int_0^{2\pi} \frac{d\phi}{2\pi} \sin^2\phi(-4(1+e\cos\phi)^2\cos\phi + 2e(3\cos^2\phi
$$

$$
- 1 + 2e\cos^3\phi))(8\cos\phi + 6e\cos^2\phi + e)
$$

$$
= -\frac{32}{5} \frac{G^{\frac{7}{2}} \mu^2 m^{\frac{5}{2}}}{c^5 a^{\frac{7}{2}}} \frac{1}{(1-e^2)^2} \left(1 + \frac{7}{8}e^2\right). \tag{3.196}
$$

Using Eqs. (3.135) and (3.138), we can rewrite Eqs. (3.196) and (3.160) as the time evolution of the semi-major axis a and of the eccentricity e in the following form

$$
\frac{da}{dt} = -\frac{64}{5} \frac{G^3 \mu m^2}{c^5 a^3} \frac{1}{(1-e^2)^{\frac{7}{2}}} \left(1 + \frac{73}{24}e^2 + \frac{37}{96}e^4\right), \tag{3.197}
$$

$$
\frac{de}{dt} = -\frac{304}{15} \frac{G^3 \mu m^2}{c^5 a^4} \frac{e}{(1-e^2)^{\frac{5}{2}}} \left(1 + \frac{121}{304}e^2\right). \tag{3.198}
$$

If $e = 0$, i.e., a circular orbit, then from Eq. (3.198) we get $\frac{de}{dt} = 0$: a circular orbit does not change. Equation (3.198) tells us that for $e > 0$, $\frac{de}{dt}$ is negative and thus an elliptic orbit becomes more and more circular due to gravitational waves emission. Note that Eqs. (3.197) and (3.198) have to be integrated numerically. Combining them, one gets

$$
\frac{da}{de} = \frac{12}{19} a \frac{1 + \frac{73}{24}e^2 + \frac{37}{96}e^4}{e(1-e^2)(1 + \frac{121}{304}e^2)}, \tag{3.199}
$$

which can be integrated analytically, leading to

$$
a(e) = c_0 \frac{e^{\frac{12}{19}}}{1-e^2} \left(1 + \frac{121}{304}e^2\right)^{\frac{870}{2299}} \equiv c_0 g(e), \tag{3.200}
$$

c_0 is determined by the initial condition $a = a_0$ when $e = e_0$, i.e., $a_0 = c_0 g(e_0)$, then $a(e) = \frac{a_0}{g(e_0)} g(e)$.

3.4.6 The PSR 1913+16 System

The Hulse–Taylor binary system is a famous example of the confirmation of the gravitational wave theory. Hulse and Taylor were awarded the 1993 Nobel Prize for the discovery and measurement of this system, which consists of two neutron stars, one of which is a pulsar whose radio signals have been observed for many years. From the phase shift of the pulsar one can infer the orbital parameters

$$\tau = 0.06 \text{ s},$$
$$T = 27{,}906.980894 \pm 0.000002 \text{ s},$$
$$M_1 = (1.442 \pm 0.003) M_\odot,$$
$$M_2 = (1.386 \pm 0.003) M_\odot,$$

where T is the orbital period of the binary system, M_1 and M_2 the masses of the two components. Because of these very accurate pulse times τ, very precise astrophysical measurements can be done with this system. Let us first assume a circular orbit for an order of magnitude estimate. We obtain for the spiral time (see Eq. 3.124)

$$t_{\text{spir.}} \sim 10^9 \text{ yr.} \tag{3.201}$$

From Kepler's law we have $T^2 \propto r^3$ such that

$$2\frac{\mathrm{d}T}{T} = 3\frac{\mathrm{d}r}{r}. \tag{3.202}$$

Furthermore Eq. (3.125) implies that

$$\left(\frac{\mathrm{d}r}{\mathrm{d}t}\right)_0 = -\frac{1}{4}\frac{r(0)}{t_{\text{spir.}}}. \tag{3.203}$$

Therefore

$$\frac{\mathrm{d}T}{\mathrm{d}t} = \frac{3}{2}\frac{\mathrm{d}r}{\mathrm{d}t}\frac{T}{r} = -\frac{3}{8}\frac{T}{t_{\text{spir.}}} \sim 10^{-12}. \tag{3.204}$$

Through the phase shift of the pulses one can measure the variation in the orbital time and finds experimentally

$$\frac{\mathrm{d}T}{\mathrm{d}t} = -(2.43 \pm 0.03) \times 10^{-12}, \tag{3.205}$$

which is indeed of the same order of magnitude as the above simplified GR estimate. As next we take into account the actual elliptical orbit and the formulas derived in the previous section. In particular using Eq. (3.165) with the definition for $f(e)$ given in Eq. (3.161) and using Kepler's third law

$$T = \frac{2\pi a^{3/2}}{(M_1 + M_2)^{1/2}}, \tag{3.206}$$

we get

$$\left\langle \frac{\dot{T}}{T} \right\rangle = -\frac{96}{5c^5} \frac{M_1 M_2}{(T/2\pi)^{8/5}(M_1 + M_2)^{1/3}} f(e). \tag{3.207}$$

Plugging in the measured ellipticity of $e \approx 0.617$, one gets

$$\dot{T}_{\text{theoretical}} = (-2.40243 \pm 0.00005) \times 10^{-12} \qquad (3.208)$$

which is in perfect agreement with the measured data. This is, indeed, indirect evidence for the existence of gravitational waves.

3.5 Waveform

Let $\hat{k}_i = \frac{k_i}{k} = n_i$, $i = 1, 2, 3$ and introduce the projector tensor

$$P_{ij}(\vec{n}) = \delta_{ij} - n_i n_j, \qquad (3.209)$$

which is symmetric. It transforms as $n^i P_{ij}(\vec{n}) = 0$ and its trace is $P_{ii} = 2$. With the help of P_{ij} we can construct $\Lambda_{ij,kl}$ as given by Eq. (3.92) as follows

$$\Lambda_{ij,kl} = P_{ik} P_{jl} - \frac{1}{2} P_{ij} P_{kl}. \qquad (3.210)$$

It is transverse on all indices

$$n^i \Lambda_{ij,kl} = n^j \Lambda_{ij,kl} = 0, \qquad (3.211)$$

and it is traceless with respect to the (ij) and (kl) indices

$$\Lambda_{ii,kl} = \Lambda_{ij,kk} = 0. \qquad (3.212)$$

Choosing $\vec{n}$ in the z direction, the transverse, traceless gauge (TT gauge) takes the following form

$$h_{ij}^{TT}(t, z) = \begin{bmatrix} h_+ & h_\times & 0 \\ h_\times & -h_+ & 0 \\ 0 & 0 & 0 \end{bmatrix}_{ij} \cos\left(\omega\left(t - \frac{z}{c}\right)\right). \qquad (3.213)$$

Using $\Lambda_{ij,kl}$ as defined in Eq. (3.92) where $\hat{k}_i = n_i$, it can be shown that the gravitational waves are given in terms of the spatial components h_{ij} of $h_{\mu\nu}$ as

$$h_{ij}^{TT} = \Lambda_{ij,kl} h_{kl}, \qquad (3.214)$$

where the summation over the k and l indices is implied. In the TT gauge, we have $T = 0$, indeed from Eq. (2.144) we see that it we take the trace, we get

$$0 = h^\mu{}_\mu \sim S^\mu{}_\mu = T^\mu{}_\mu - \frac{T}{2}\delta^\mu{}_\mu = -T. \tag{3.215}$$

Therefore we get with Eqs. (3.214), (2.144) and $T = 0$

$$h^{TT}{}_{ij}(t, \vec{r}) = -\frac{4G}{c^4}\Lambda_{ij,kl}\int d^3r'\frac{T_{kl}(\vec{r}', t - \frac{|\vec{r}-\vec{r}'|}{c})}{|\vec{r} - \vec{r}'|}. \tag{3.216}$$

As we are interested in the value of h^{TT}_{ij} at large distances from the source, we take $r \to \infty$. To leading order we get

$$\frac{T_{kl}}{|\vec{r} - \vec{r}'|} \sim \frac{1}{r}T_{kl}(\vec{r}', t - \frac{r}{c} + \frac{\vec{r}'\hat{\vec{n}}}{c}), \tag{3.217}$$

since

$$\frac{|\vec{r} - \vec{r}'|}{c} = \frac{r}{c} - \frac{\vec{r}'\hat{\vec{n}}}{c} + O\left(\frac{r_0^2}{r}\right), \tag{3.218}$$

where r_0 is the size of the source. Furthermore performing a Taylor expansion

$$T_{kl}(\vec{r}', t - \frac{r}{c} + \frac{\vec{r}'\hat{\vec{n}}}{c}) \simeq T_{kl}(\vec{r}', t - \frac{r}{c}) + \frac{r'^i n^i}{c}\partial_0 T_{kl} + \frac{1}{2c^2}r'^i r'^j n^i n^j \partial_0^2 T_{kl} + \cdots, \tag{3.219}$$

where all derivatives are evaluated at the point $(\vec{r}', t - \frac{r}{c})$. We define the momenta of the stress tensor T^{ij} as following

$$S^{ij}(t_{\text{ret}}) = \int d^3x\, T^{ij}(t_{\text{ret}}, \vec{x}), \tag{3.220}$$

$$S^{ij,k}(t_{\text{ret}}) = \int d^3x\, T^{ij}(t_{\text{ret}}, \vec{x})x^k, \tag{3.221}$$

$$S^{ij,kl}(t_{\text{ret}}) = \int d^3x\, T^{ij}(t_{\text{ret}}, \vec{x})x^k x^l. \tag{3.222}$$

In this notation the comma separates the spatial indices of T^{ij} from the indices due to x^i. Furthermore, note that the separated indices from T^{ij} with the ones from x^k, x^l can be interchanged, i.e., $i \leftrightarrow j$ and $k \leftrightarrow l$. Inserting the expression (3.219) into Eq. (3.217), we get

$$h^{TT}_{ij}(\vec{r}, t) = \frac{1}{r}\frac{4G}{c^4}\Lambda_{ij,kl}(\hat{\vec{n}})\left(S^{kl} + \frac{1}{c}n_m \dot{S}^{kl,m} + \frac{1}{2c^2}n_m n_p \ddot{S}^{kl,mp} + \cdots\right), \tag{3.223}$$

where S^{kl} and its derivatives are evaluated at the retarded time $t_{\text{ret}} = t - \frac{r}{c}$. The physical meaning of the various terms in the expansion can be seen more clearly once we eliminate the momenta of T^{ij} in terms of the momenta of the energy density

T^{00}, and of the linear momentum $\frac{T^{0i}}{c}$. The momenta of $\frac{T^{00}}{c^2}$ are as follows

$$M(t) = \frac{1}{c^2} \int d^3x \, T^{00}(t, \vec{x}), \tag{3.224}$$

$$M^i(t) = \frac{1}{c^2} \int d^3x \, T^{00}(t, \vec{x}) x^i, \text{ etc.,} \tag{3.225}$$

and the momenta of the momentum density $\frac{T^{0i}}{c}$ by

$$P^i(t) = \frac{1}{c} \int d^3x \, T^{0i}(t, \vec{x}), \tag{3.226}$$

$$P^{i,j}(t) = \frac{1}{c} \int d^3x \, T^{0i}(t, \vec{x}) x^j, \text{ etc..} \tag{3.227}$$

In the flat spacetime approximation $T^{\mu\nu}$ satisfies $\partial_\mu T^{\mu\nu} = 0$. Thus, $\partial_0 T^{00} = -\partial_i T^{0i}$ and recalling that $\dot{M} = c \partial_0 M$, it follows

$$c\dot{M} = \int_V d^3x \, \partial_0 T^{00} = -\int_V d^3x \, \partial_i T^{0i} = -\int_{\partial V} dS^i \, T^{0i} = 0, \tag{3.228}$$

where in the last equality we used that T^{0i} vanishes on the boundary ∂V, as we have taken the volume V larger than the volume of the source. Similarly, neglecting mass loss of the source, we obtain

$$c\dot{M}^i = \int_V d^3x \, x^i \partial_0 T^{00} = -\int_V d^3x \, x^i \partial_j T^{0j} = \int d^3x \, (\partial_j x^i) T^{0j}$$

$$= \int d^3x \, \delta^i_j T^{0j} = \int d^3x \, T^{0i} = c P^i. \tag{3.229}$$

One can show in the same ways that the following relations hold

$$\dot{M} = 0, \quad \dot{M}^i = P^i, \quad \dot{M}^{ij} = P^{i,j} + P^{j,i} \quad \dot{M}^{ijk} = P^{i,jk} + P^{j,ki} + P^{k,ij}, \tag{3.230}$$

where $\dot{M} = 0$ is the mass conservation and

$$\dot{P}^i = 0, \quad \dot{P}^{i,j} = S^{ij}, \quad \dot{P}^{i,jk} = S^{ij,k} + S^{ik,j}, \tag{3.231}$$

where $\dot{P}^i = 0$ is the momentum conservation of the source. Using the above relations, one gets

$$S^{ij} = \frac{1}{2} \ddot{M}^{ij}, \quad \dddot{M}^{ijk} = 2(\dot{S}^{ij,k} + \dot{S}^{ik,j} + \dot{S}^{jk,i}), \quad \ddot{P}^{i,jk} = \dot{S}^{ij,k} + \dot{S}^{ik,j}, \tag{3.232}$$

$$\Rightarrow \dot{S}^{ij,k} = \frac{1}{6}\dddot{M}^{ijk} + \frac{1}{3}(\ddot{P}^{i,jk} + \ddot{P}^{j,ik} - 2\ddot{P}^{k,ij}). \tag{3.233}$$

These relations can be inserted into Eq. (3.223). One can proceed similarly with higher-order terms ($\sim \dddot{S}^{kl,mp}$). To leading order, quadrupole radiation, Eq. (3.223) becomes

$$h_{ij}^{TT} = \frac{1}{r}\frac{2G}{c^4}\Lambda_{ij,kl}(\hat{n})\ddot{M}^{kl}\left(t - \frac{r}{c}\right), \tag{3.234}$$

with

$$M^{kl} = (M^{kl} - \frac{1}{3}\delta^{kl}M_{ii}) + \frac{1}{3}\delta^{kl}M_{ii}, \tag{3.235}$$

where M_{ii} is the trace of M_{ij}. The first term is traceless since $\Lambda_{ij,kl}$ gives zero when contracted with δ_{kl}, only the first term contributes. The quadrupole moment is then as follows

$$Q^{ij} = M^{ij} - \frac{1}{3}\delta^{ij}M_{kl} = \int d^3x \left(x^i x^j - \frac{1}{3}r^2\delta^{ij}\right)\rho(t, \vec{x}), \tag{3.236}$$

then Eq. (3.234) becomes

$$h_{ij}^{TT}(t, \vec{r}) = \frac{1}{r}\frac{2G}{c^4}\Lambda_{ij,kl}(\hat{n})\ddot{Q}_{kl}\left(t - \frac{r}{c}\right). \tag{3.237}$$

Assume a wave traveling along the z direction, thus $\hat{n} = \hat{z}$, then it can be shown that

$$\Lambda_{ij,kl}\ddot{M}_{kl} = \begin{bmatrix} (\ddot{M}_{11} - \ddot{M}_{22})/2 & \ddot{M}_{12} & 0 \\ \ddot{M}_{21} & -(\ddot{M}_{11} - \ddot{M}_{22})/2 & 0 \\ 0 & 0 & 0 \end{bmatrix}_{ij}, \tag{3.238}$$

and thus

$$h_+ = \frac{1}{r}\frac{G}{c^4}(\ddot{M}_{11} - \ddot{M}_{22}), \tag{3.239}$$

$$h_\times = \frac{2}{r}\frac{G}{c^4}\ddot{M}_{12}, \tag{3.240}$$

where the right-hand side is computed at the retarded time $t_{\text{ret}} = t - \frac{r}{c}$.

For a generic direction $n_i = (\sin\theta\sin\phi, \sin\theta\cos\phi, \cos\theta)$, and the components n_i and n_i' (for instance n_i' in z-axis) are related by the rotation matrix $R_{ij}: n_i = R_{ij}n_j'$ with

$$R = \begin{bmatrix} \cos\phi & \sin\phi & 0 \\ -\sin\phi & \cos\phi & 0 \\ 0 & 0 & 1 \end{bmatrix}\begin{bmatrix} 1 & 0 & 0 \\ 0 & \cos\theta & \sin\theta \\ 0 & -\sin\theta & \cos\theta \end{bmatrix} \tag{3.241}$$

and $M_{ij} = R_{ik}R_{jl}M_{kl}'$. We get then

$$h_+(t, \theta, \phi) = \frac{1}{r}\frac{G}{c^4}\Big(\ddot{M}_{11}(\cos^2\phi - \sin^2\phi \cos^2\theta) + \ddot{M}_{22}(\sin^2\phi - \cos^2\phi \cos^2\theta) +$$

$$- \ddot{M}_{33}\sin^2\theta - \ddot{M}_{12}\sin 2\phi(1 + \cos^2\theta) + \ddot{M}_{13}\sin\phi \sin 2\theta$$

$$+ \ddot{M}_{23}\cos\phi \sin 2\theta \Big), \tag{3.242}$$

$$h_\times(t, \theta, \phi) = \frac{1}{r}\frac{G}{c^4}\Big((\ddot{M}_{11} - \ddot{M}_{22})\sin 2\phi \cos\theta + 2\ddot{M}_{12}\cos 2\phi \cos\theta$$

$$- 2\ddot{M}_{13}\cos\phi \sin\theta + 2\ddot{M}_{23}\sin\phi \sin\theta \Big), \tag{3.243}$$

and $\ddot{M}_{ij}$ is evaluated at $t - \frac{r}{c}$. With this equation one can compute the angular distribution of the quadrupole radiation.

3.5.1 Quadrupole Radiation from a Mass in Circular Orbit

We consider a binary system with masses m_1 and m_2 on a circular orbit of radius R. We neglect any back-reaction on the motion due to gravitational waves emission. We chose our coordinate system such that the orbit lies in the (x, y) plane and it is given by

$$x_0(t) = R \cos\left(\omega_s t + \frac{\pi}{2}\right), \tag{3.244}$$

$$y_0(t) = R \sin\left(\omega_s t + \frac{\pi}{2}\right), \tag{3.245}$$

$$z_0(t) = 0, \tag{3.246}$$

where the phase $\frac{\pi}{2}$ is a useful choice of the origin of time. We use the reduced mass μ of the system to express the second mass moment

$$M^{ij} = \mu x_0^i(t) x_0^j(t), \tag{3.247}$$

in the center of mass coordinate system. Using the trigonometric identity $\cos 2\alpha = 2\cos^2\alpha - 1$ and the coordinates given in Eq. (3.246), we get

$$M_{11} = \mu R^2 \frac{1 - \cos(2\omega_s t)}{2}, \tag{3.248}$$

$$M_{22} = \mu R^2 \frac{1 + \cos(2\omega_s t)}{2}, \tag{3.249}$$

$$M_{12} = -\frac{1}{2}\mu R^2 \sin 2\omega_s t, \tag{3.250}$$

while the other components vanish and we also compute

$$\ddot{M}_{11} = 2\mu R^2 \omega_s^2 \cos(2\omega_s t), \tag{3.251}$$

$$\ddot{M}_{12} = 2\mu R^2 \omega_s^2 \sin(2\omega_s t), \tag{3.252}$$

$$\ddot{M}_{11} = -\ddot{M}_{22}. \tag{3.253}$$

Inserting these relations into Eqs. (3.242) and (3.243), we get

$$h_+(t, \theta, \phi) = \frac{1}{r} \frac{4G\mu\omega_s^2 R^2}{c^4} \left(\frac{1 + \cos^2 \theta}{2} \right) \cos(2\omega_s t_{\text{ret}} + 2\phi), \tag{3.254}$$

$$h_\times(t, \theta, \phi) = \frac{1}{r} \frac{4G\mu\omega_s^2 R^2}{c^4} \cos\theta \sin(2\omega_s t_{\text{ret}} + 2\phi), \tag{3.255}$$

where the orbital frequency ω_s is related to the orbital radius R by $v = \omega_s R$ and $\frac{v^2}{R} = \frac{Gm}{R^2}$, where $m = m_1 + m_2$ and thus $\omega_s = \frac{Gm}{R^3}$. Note that the quadrupole radiation given in Eqs. (3.254) and (3.255) has twice the frequency ω_s of the source.

We now introduce the *chirp mass*

$$M_c = \mu^{\frac{3}{5}} m^{\frac{2}{5}} = \frac{(m_1 m_2)^{\frac{3}{5}}}{(m_1 + m_2)^{\frac{1}{5}}}, \tag{3.256}$$

moreover we define $f_{GW} := \frac{\omega_{GW}}{2\pi}$ where $\omega_{GW} = 2\omega_s$. Then Eqs. (3.254) and (3.255) become

$$h_+(t, \theta, \phi) = \frac{4}{r} \left(\frac{GM_c}{c^2} \right)^{\frac{5}{3}} \left(\frac{\pi f_{GW}}{c} \right)^{\frac{2}{3}} \frac{1 + \cos^2 \theta}{2} \cos(2\pi f_{GW} t_{\text{ret}} + 2\phi), \tag{3.257}$$

$$h_\times(t, \theta, \phi) = \frac{4}{r} \left(\frac{GM_c}{c^2} \right)^{\frac{5}{3}} \left(\frac{\pi f_{GW}}{c} \right)^{\frac{2}{3}} \cos\theta \sin(2\pi f_{GW} t_{\text{ret}} + 2\phi). \tag{3.258}$$

From Eqs. (3.124) and (3.125) it follows that the typical inspiral time, i.e., time to coalescence, is

$$t_{\text{spir}} = \frac{5}{256} \frac{c^5 R_0^4}{G^3 m^2 \mu}. \tag{3.259}$$

Expressing the initial radius R_0 in terms of the initial orbital period $T_0 = \frac{2\pi}{\omega_s (t_{\text{spir}})}$ through Kepler's third law $R_0^3 = Gm(T_0/2\pi)^2$, we find

$$t_{\text{spir}} \simeq 9.8 \times 10^6 \text{ yr} \left(\frac{T_0}{1 \text{ hour}} \right)^{\frac{8}{3}} \left(\frac{M_\odot}{m} \right)^{\frac{2}{3}} \left(\frac{M_\odot}{\mu} \right). \tag{3.260}$$

Thus, for masses $\sim M_\odot$ in circular orbits: Only binaries which had at formation an initial period of less than about one day can have coalesced by emission of gravitational waves within the present age of galaxies.

A particle that moves on a quasi-circular orbit in the (x, y) plane with a radius $R = R(t)$ and angular velocity $\omega_s = \omega_s(t)$ has Cartesian coordinates $x(t) = R(t)\cos(\Phi(t)/2)$ and $y(t) = R(t)\sin(\Phi(t)/2)$, with

$$\Phi(t) = 2\int_{t_0}^{t} dt'\omega_s(t') = \int_{t_0}^{t} dt'\omega_{GW}(t'). \tag{3.261}$$

When computing the gravitational waves using the quadrupole approximation in the time derivatives of the mass moments we get additional terms. This implies the following:

- in the argument of the trigonometric functions, $\omega_{GW}t$ has to be replaced by $\Phi(t)$;
- in the factors in front of the trigonometric functions, instead of ω_{GW} or f_{GW} one has to insert $\omega_{GW}(t)$ or $f_{GW}(t)$;
- one should also include the contributions coming from the derivatives of $R(t)$ and $\omega_{GW}(t)$, however $\dot{R}$ is negligible as long as $\dot{\omega}_s \ll \omega_s^2$. In the following we shall assume that this condition is fulfilled and thus neglect those terms.

Eqs. (3.257) and (3.258) become then

$$h_+(t) = \frac{4}{r}\left(\frac{GM_c}{c^2}\right)^{\frac{5}{3}}\left(\frac{\pi f_{GW}(t_{\text{ret}})}{c}\right)^{\frac{2}{3}}\left(\frac{1+\cos^2\theta}{2}\right)\cos\Phi(t_{\text{ret}}), \tag{3.262}$$

$$h_\times(t) = \frac{4}{r}\left(\frac{GM_c}{c^2}\right)^{\frac{5}{3}}\left(\frac{\pi f_{GW}(t_{\text{ret}})}{c}\right)^{\frac{2}{3}}\cos\theta\sin\Phi(t_{\text{ret}}), \tag{3.263}$$

where $f_{GW} \to f_{GW}(t_{\text{ret}})$ and $2\pi f_{GW}t_{\text{ret}} + 2\phi \to \Phi(t_{\text{ret}})$. In the coalescence phase f_{GW} increases to $f_{GW} \sim \frac{1}{T} \to \infty$ (T orbital period), however when the black holes are very closed all the approximations we used break down. The dependence of h_+ or $h_\times$ on t is shown in Fig. 3.4.

Fig. 3.4 Time evolution of the gravitational wave amplitude in the inspiral phase of a binary system

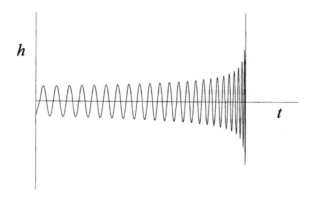

3.6 The Post-Newtonian Approximation

Astrophysical sources of gravitational radiation are held together by gravitational forces. For a self-gravitating system of mass m we have

$$\left(\frac{v}{c}\right)^2 \sim \frac{R_S}{d},\tag{3.264}$$

where R_S is the Schwarzschild radius and d is a typical size of the system. This relation follows immediately from $v^2 \approx Gm/r \approx Gm/d$. We note that R_S/d is a measure of the strength of the gravitational field close to the source. As soon as we consider v/c corrections to the orbital motion, for consistency we have to consider also $(R_S/d)^{1/2}$ corrections to the metric (corrections to the flat background).

Moderately relativistic systems require a **post-Newtonian** treatment. The assumptions that we will use are

- The systems under consideration are moving slowly, weakly self-gravitating systems such that an expansion in v/c or $(R_S/d)^{1/2}$ is possible.
- The energy-momentum tensor $T_{\mu\nu}$ has a spatially compact support ($T_{\mu\nu}(r) = 0$ for $r \geq d$).

If ω_S is a typical frequency of the system, then typical velocities are $v \approx \omega_S d$. As we saw before, the frequency of the radiation of the emitted gravitational radiation is $\omega = 2\omega_S \approx 2v/d$. In non-relativistic systems we have $v \ll c$, thus $c/v \gg 1$ and the wavelength of the emitted radiation satisfies $\lambda = c/\omega \sim cd/v \gg d$.

In analogy to the electromagnetic case, for non-relativistic sources it is convenient to distinguish between:

- *near field* regime ($r, d \ll \lambda$) where retardation is negligible and potentials are static,
- *far field* regime ($r \gg \lambda$) where retardation is crucial and we have waves.

The small parameter in powers of which we will perform an expansion is[6]

$$\varepsilon \sim \left(\frac{R_S}{d}\right)^{1/2} \sim \frac{v}{c}.\tag{3.265}$$

We require that

$$\frac{|T^{ij}|}{T^{00}} \sim \mathcal{O}(\varepsilon^2).\tag{3.266}$$

For a fluid with pressure p and energy density ρ, we thus have

$$\frac{p}{\rho} \sim \varepsilon^2.\tag{3.267}$$

---

[6] Note that some references also use the convention $\varepsilon \sim (v/c)^2$.

We expand the metric and the energy-momentum tensor in the near-field regime in powers of ε.

As long as the emission of radiation is neglected, a classical system subject to conservative forces is invariant under time reversal. Under time reversal, g_{00} and g_{ij} are even (i.e., there appear even powers of v and thus of ε, as well) while g_{0i} is odd (i.e., only odd powers of v and thus of ε appear). By inspection of Einstein's field equations, one finds that in order to work consistently to a given order ε, if we expand g_{00} up to $\mathcal{O}(\varepsilon^n)$, then we have to expand g_{0i} up to $\mathcal{O}(\varepsilon^{n-1})$ and g_{ij} up to $\mathcal{O}(\varepsilon^{n-2})$.

The metric is expanded as follows:

$$
\begin{aligned}
g_{00} &= -1 + {}^{(2)}g_{00} + {}^{(4)}g_{00} + {}^{(6)}g_{00} + \cdots \\
g_{0i} &= \qquad\quad {}^{(3)}g_{0i} + {}^{(5)}g_{0i} + \cdots \\
g_{ij} &= \delta_{ij} + {}^{(2)}g_{ij} + {}^{(4)}g_{ij} + \cdots ,
\end{aligned}
\tag{3.268}
$$

where ${}^{(n)}g_{\mu\nu}$ denotes terms of $\mathcal{O}(\varepsilon^n)$. Similarly, for the energy-momentum tensor:

$$
\begin{aligned}
T^{00} &= {}^{(0)}T^{00} + {}^{(2)}T^{00} + \cdots \\
T^{0i} &= {}^{(1)}T^{0i} + {}^{(3)}T^{0i} + \cdots \\
T^{ij} &= {}^{(2)}T^{ij} + {}^{(4)}T^{ij} + \cdots
\end{aligned}
\tag{3.269}
$$

We now want to insert these expansions into Einstein's field equations and equate terms of the same order in ε. Considering $v \ll c$, the time derivatives of the metric are smaller than the spatial derivatives by $\mathcal{O}(\varepsilon)$:

$$
\frac{\partial}{\partial t} = \mathcal{O}(v)\frac{\partial}{\partial x} \qquad \text{or} \qquad \partial_0 \sim \mathcal{O}(\varepsilon)\partial_i,
\tag{3.270}
$$

where we used that $\partial_0 = \frac{1}{c}\partial_t$. The d'Alembert operator applied to the metric to lowest order becomes the Laplacian:

$$
\left(-\frac{1}{c^2}\frac{\partial^2}{\partial t^2} + \Delta\right) = \left(\mathcal{O}(\varepsilon^2) + 1\right)\Delta.
\tag{3.271}
$$

Thus retardation effects are small corrections.

Consequently, we also have to expand the geodesic equation (Eq. 1.43)

$$
\frac{d^2 x^i}{d\tau^2} = -\Gamma^i_{\mu\nu}\frac{dx^\mu}{d\tau}\frac{dx^\nu}{d\tau}.
\tag{3.272}
$$

In Newtonian limit we have $g_{00} = -1 + {}^{(2)}g_{00}$, $g_{0i} = 0$ and $g_{ij} = \delta_{ij}$. It thus follows that the terms ${}^{(4)}g_{00}$, ${}^{(3)}g_{0i}$ and ${}^{(2)}g_{ij}$ give the first post-Newtonian order for which we use the notation 1PN. The terms ${}^{(6)}g_{00}$, ${}^{(5)}g_{0i}$, ${}^{(4)}g_{ij}$ give the 2PN approximation and so on.

3.6.1 The 1PN Approximation

It is useful to choose a simplifying gauge condition right from the beginning. A convenient choice is the *de Donder gauge condition*

$$\partial_\mu(\sqrt{-g}\,g^{\mu\nu}) = 0. \tag{3.273}$$

(this is a harmonic gauge condition, i.e., the coordinate functions satisfy the d'Alembert equation).

The next step is to insert Eqs. (3.268) and (3.269) into Einstein's equations (together with 3.270 and 3.273). We skip the explicit computations and give only the results (for a detailed derivation, see, e.g., [11, 12]). One finds for $^{(2)}g_{00}$ the Newtonian equation

$$\Delta\left[^{(2)}g_{00}\right] = -\frac{8\pi G}{c^4}\,^{(0)}T^{00}, \tag{3.274}$$

while the 1PN correction to the metric yields

$$\Delta\left[^{(2)}g_{ij}\right] = -\frac{8\pi G}{c^4}\delta_{ij}\,^{(0)}T^{00}, \tag{3.275}$$

$$\Delta\left[^{(3)}g_{0i}\right] = \frac{16\pi G}{c^4}\,^{(1)}T^{0i}, \tag{3.276}$$

$$\Delta\left[^{(4)}g_{00}\right] = \partial_0^2\left[^{(2)}g_{00}\right] + {}^{(2)}g_{ij}\partial_i\partial_j\left[^{(2)}g_{00}\right] - \partial_i\left[^{(2)}g_{00}\right]\partial_i\left[^{(2)}g_{00}\right]$$
$$- \frac{8\pi G}{c^4}\left[^{(2)}T^{00} + {}^{(2)}T^{ii} - 2^{(2)}g_{00}\,^{(0)}T^{00}\right], \tag{3.277}$$

where $\Delta = \delta^{ij}\partial_i\partial_j$ and the sum over repeated spatial indices is performed with δ^{ij}.

The solution of (3.274) with the boundary condition that the metric is asymptotically flat, is

$$^{(2)}g_{00} = -2\phi \quad \text{with } \phi(t,\vec{x}) = -\frac{G}{c^4}\int d^3x'\frac{^{(0)}T^{00}(\vec{x}',t)}{|\vec{x}-\vec{x}'|}, \tag{3.278}$$

where the Newtonian potential is $U = -c^2\phi$. Similarly Eqs. (3.275) and (3.276) are solved by

$$^{(2)}g_{ij} = -2\phi\delta_{ij} \tag{3.279}$$

$$^{(3)}g_{0i} = \xi_i, \tag{3.280}$$

where we defined

$$\xi_i(\vec{x},t) = -\frac{4G}{c^4}\int d^3x'\frac{^{(1)}T^{0i}(\vec{x}',t)}{|\vec{x}-\vec{x}'|}. \tag{3.281}$$

In order to solve (3.277), we replace $^{(2)}g_{00}$ on the right-hand side by -2ϕ and $^{(2)}g_{ij}$ by $-2\phi\delta_{ij}$. Furthermore we use the identity

$$(\vec{\nabla}\phi)^2 = \partial_i\phi\partial_i\phi = \frac{1}{2}\Delta(\phi^2) - \phi\Delta\phi. \tag{3.282}$$

and we introduce a new potential ψ such that

$$^{(4)}g_{00} = -2(\phi^2 + \psi). \tag{3.283}$$

Eq. (3.277) then reads

$$\Delta\psi = \partial_0^2\phi + \frac{4\pi G}{c^4}\left(^{(2)}T^{00} + {}^{(2)}T^{ii}\right). \tag{3.284}$$

Using the boundary condition that ψ vanishes at spatial infinity, ψ can be written as

$$\psi(\vec{x}, t) = -\int \frac{d^3x'}{|\vec{x} - \vec{x}'|}\left(\frac{1}{4\pi}\partial_0^2\phi + \frac{G}{c^4}\left[^{(2)}T^{00}(\vec{x}', t) + {}^{(2)}T^{ii}(\vec{x}', t)\right]\right). \tag{3.285}$$

Notice that ϕ and ξ_i are not independent due to the gauge condition (3.273) which imposes the constraint

$$4\partial_0\phi + \vec{\nabla} \cdot \vec{\xi} = 0. \tag{3.286}$$

From Eqs. (3.278) and (3.281) one can see that this condition is indeed satisfied due to energy-momentum conservation at the 1PN order (since $T^{\mu\nu}$ is covariantly conserved in the exact solution, it has to be conserved at all post-Newtonian orders independently).

We observe that ϕ, ψ, ξ_i are instantaneous potentials. Our order of approximation is thus insensitive to retardation effects. Note also that g_{00} can be expressed very simply as

$$g_{00} = -e^{-2V/c^2} + \mathcal{O}(\varepsilon^6), \tag{3.287}$$

where $V = -c^2(\phi + \psi)$. This follows immediately if we expand the exponential and write (3.287) as

$$\begin{aligned}
g_{00} &= -1 + \frac{2V}{c^2} - \frac{2V^2}{c^4} + \mathcal{O}\left(\frac{1}{c^6}\right) \\
&= -1 - 2(\phi + \psi) - 2(\phi + \psi)^2 + \mathcal{O}(\varepsilon^6) \\
&= -1 - 2\phi - 2(\phi^2 + \psi) - 2(\psi^2 + 2\phi\psi) + \mathcal{O}(\varepsilon^6).
\end{aligned} \tag{3.288}$$

Using that $\phi = \mathcal{O}(\varepsilon^2)$ and $\psi = \mathcal{O}(\varepsilon^4)$, we see that this is just

$$g_{00} = -1 + {}^{(2)}g_{00} + {}^{(4)}g_{00} + \mathcal{O}(\varepsilon^6). \tag{3.289}$$

Putting together (3.274) (with $^{(2)}g_{00} = -2\phi$) and (3.284), we have

$$\Delta(\phi + \psi) = \partial_0^2\phi + \frac{4\pi G}{c^4}\left[^{(0)}T^{00} + {}^{(2)}T^{00} + {}^{(2)}T^{ii}\right].\qquad(3.290)$$

To this order we can set $\partial_0^2\phi = \partial_0^2(\phi + \psi)$ and replace Δ by $\Box$. We then obtain

$$\Box V = -\frac{4\pi G}{c^2}\left[T^{00} + T^{ii}\right] \equiv -\frac{4\pi G}{c^2}\sigma,\qquad(3.291)$$

where we replaced $^{(0)}T^{00} + {}^{(2)}T^{00} \to T^{00}$ and $^{(2)}T^{ii} \to T^{ii}$. The solution of Eq. (3.291) is given by a retarded integral

$$V(\vec{x}, t) = G\int d^3x' \frac{1}{|\vec{x} - \vec{x}'|}\left(T^{00} + T^{ii}\right)\left(t - \frac{|\vec{x} - \vec{x}'|}{c}, \vec{x}'\right)\qquad(3.292)$$

and similarly for ξ_i given by

$$\xi_i(\vec{x}, t) = G\int d^3x' \frac{1}{|\vec{x} - \vec{x}'|}T^{0i}\left(t - \frac{|\vec{x} - \vec{x}'|}{c}, \vec{x}'\right).\qquad(3.293)$$

To summarize, in harmonic coordinates the 1PN solution can be written in terms of two functions V and ξ_i in the following way:

$$\boxed{\begin{aligned}
g_{00} &= -1 + \frac{2}{c^2}V - \frac{2}{c^4}V^2 + \mathcal{O}\left(\frac{1}{c^6}\right),\\[2mm]
g_{0i} &= -\frac{4}{c^3}\xi_i + \mathcal{O}\left(\frac{1}{c^5}\right),\\[2mm]
g_{ij} &= \delta_{ij}\left(1 + \frac{2}{c^2}V\right) + \mathcal{O}\left(\frac{1}{c^4}\right).
\end{aligned}}\qquad(3.294)$$

3.6.2 Equations of Motion of Test Particles in the 1PN Metric

To get the equations of motion of a particle of mass m in the near zone, we have to solve the geodesic equations, which are obtained by applying the variational principle to the following action, which in curved background is as follows

$$\begin{aligned}
S &= -mc\int dt \left(-g_{\mu\nu}\frac{dx^\mu}{dt}\frac{dx^\nu}{dt}\right)^{1/2}\\[2mm]
&= -mc^2\int dt \left(-g_{00} - 2g_{0i}\frac{v^i}{c} - g_{ij}\frac{v^i v^j}{c^2} + \cdots\right)^{1/2}.\qquad(3.295)
\end{aligned}$$

We are interested in the equations of motion for a binary system. If we restrict ourselves to the lowest PN corrections, it is possible to treat the two masses as point-like.

In curved space the energy-momentum tensor of a set of point-like particles is given by

$$T^{\mu\nu} = \frac{1}{\sqrt{-g}} \sum_{a=1}^{N} \gamma_a m_a \frac{dx_a^\mu}{dt} \frac{dx_a^\nu}{dt} \delta^{(3)} (\vec{x} - \vec{x}_a(t)), \qquad (3.296)$$

where the masses are denoted by m_a and x_a^μ are their coordinates ($a = 1, \ldots, N$). Furthermore the following definitions hold

$$\gamma_a^{-1} = \frac{d\tau_a}{dt} = \sqrt{-g_{00} - g_{ij} v_a^i v_a^j / c^2} \quad \text{and} \quad p_a^\mu = \gamma_a m_a \frac{dx_a^\mu}{dt}. \qquad (3.297)$$

In an N-body system ($a > 2$) the metric felt by a particle b is obtained by taking the energy-momentum tensor of all other particles as a source. This amounts to replacing $\sum_a$ by $\sum_{a(\neq b)}$ in Eq. (3.296). We expand the determinant of the metric to second order and using $^{(2)}g_{00} = -2\phi$, we get

$$-g = 1 - {}^{(2)}g_{00} + \sum_{i=1}^{3} {}^{(2)}g_{ii} = 1 - 4\phi. \qquad (3.298)$$

Therefore the expansion of (3.296) gives

$$\begin{aligned}
{}^{(0)}T^{00}(\vec{x}, t) &= \sum_{a(\neq b)} m_a c^2 \delta^{(3)}(\vec{x} - \vec{x}_a(t)) \\
{}^{(2)}T^{00}(\vec{x}, t) &= \sum_{a(\neq b)} m_a \left(\frac{1}{2} v_a^2 + \phi c^2 \right) \delta^{(3)}(\vec{x} - \vec{x}_a(t)) \\
{}^{(1)}T^{0i}(\vec{x}, t) &= c \sum_{a(\neq b)} m_a v_a^i \delta^{(3)}(\vec{x} - \vec{x}_a(t)) \\
{}^{(2)}T^{ij}(\vec{x}, t) &= \sum_{a(\neq b)} m_a v_a^i v_a^j \delta^{(3)}(\vec{x} - \vec{x}_a(t)).
\end{aligned} \qquad (3.299)$$

Inserting these expressions into Eqs. (3.278), (3.281) and (3.285), one can obtain the metric in which the particle b propagates. Inserting this metric into (3.295), one can calculate its action S_b. The total action can then be obtained by summing over all particles, $S = \sum_b S_b$. Expanding the square root that appears in the integral of the action, to get the 1PN corrections one has to consistently consider all terms up to $\mathcal{O}\left((v/c)^4\right)$.

In terms of the Lagrangian, one can verify the following results for the two-body system:

$$\mathcal{L} = \mathcal{L}_0 + \frac{1}{c^2}\mathcal{L}_1 \tag{3.300}$$

$$\text{with } \mathcal{L}_0 = \frac{1}{2}m_1 v_1^2 + \frac{1}{2}m_2 v_2^2 \frac{Gm_1 m_2}{r}, \tag{3.301}$$

$$\mathcal{L}_1 = \frac{1}{8}m_1 v_1^4 + \frac{1}{8}m_2 v_2^4$$
$$+ \frac{Gm_1 m_2}{2r}\left[3(v_1^2 + v_2^2) - 7\vec{v}_1\vec{v}_2 - (\hat{r}\vec{v}_1)(\hat{r}\vec{v}_2) - \frac{G(m_1 + m_2)}{r}\right], \tag{3.302}$$

where r denotes the separation vector between the two particles, $r = |\vec{r}|$ and $\hat{r} = \frac{\vec{r}}{r}$. The Lagrangian's $\mathcal{L}_0$ and $\mathcal{L}_1$ describe the Newtonian part and the first post-Newtonian correction, respectively.

The same can be obtained for N particles (*Einstein–Infeld–Hoffmann Lagrangian*) :

$$\mathcal{L}_0 = \sum_a \frac{1}{2}m_a v_a^2 + \sum_a \sum_{b(\neq a)} \frac{Gm_a m_b}{2r_{ab}}, \tag{3.303}$$

$$\mathcal{L}_1 = \sum_a \frac{1}{8}m_a v_a^4 - \sum_a \sum_{b(\neq a)} \frac{Gm_a m_b}{4r_{ab}}\left[7\vec{v}_a\vec{v}_b + (\hat{r}_{ab}\vec{v}_a)(\hat{r}_{ab}\vec{v}_b)\right]$$
$$+ \frac{3G}{2}\sum_a \sum_{b(\neq a)} \frac{m_a m_b v_a^2}{r_{ab}} - \frac{G^2}{2}\sum_a \sum_{b(\neq a)} \sum_{c(\neq a)} \frac{m_a m_b m_c}{r_{ab} r_{ac}}, \tag{3.304}$$

where $a = 1, \ldots, N$ labels the particles, r_{ab} is the distance between particles a and b, and $\hat{r}_{ab}$ is the corresponding unit vector. From this Lagrangian one can derive the Euler–Lagrange equations of the N particle system including 1PN corrections. These equations are also called the *Einstein–Infeld–Hoffmann equations* [13]. Denoting $r_{ab} = |\vec{x}_a - \vec{x}_b|$ and $\vec{x}_{ab} = \vec{x}_a - \vec{x}_b$, one finds after lengthy calculations ($c = 1$)

$$\dot{\vec{v}}_a = -G\sum_{b\neq a} m_b \frac{\vec{x}_{ab}}{r_{ab}^3}\left[1 - 4G\sum_{c\neq a}\frac{m_c}{r_{ac}} + G\sum_{c\neq b}m_c\left(-\frac{1}{r_{bc}} + \frac{\vec{x}_{ab}\vec{x}_{bc}}{2r_{bc}^3}\right)\right.$$
$$\left. + v_a^2 - 4\vec{v}_a\vec{v}_b + 2v_b^2 - \frac{3}{2}\left(\frac{\vec{v}_b\vec{x}_{ab}}{r_{ab}}\right)^2\right]$$
$$- \frac{7}{2}G^2\sum_{b\neq a}\left(\frac{m_b}{r_{ab}}\right)\sum_{c\neq b}\frac{m_c\vec{x}_{bc}}{r_{bc}^3} + G\sum_{b\neq a}m_b\left(\frac{\vec{x}_{ab}}{r_{ab}^3}\right)(4\vec{v}_a - 3\vec{v}_b)\cdot(\vec{v}_a - \vec{v}_b). \tag{3.305}$$

3.6.3   Two Body Problem in the 1PN Approximation

The Einstein–Infeld–Hoffmann equations for the two body problem imply that the center of mass

$$\vec{X} = \frac{m_1^* \vec{x}_1 + m_2^* \vec{x}_2}{m_1^* + m_2^*} \tag{3.306}$$

with

$$m_a^* := m_a + \frac{1}{2} m_a \left(\frac{v_a}{c} \right)^2 - \frac{1}{2} \frac{m_a m_b}{r_{ab}} \frac{G}{c^2} \qquad (a, b = 1, 2, \ a \neq b) \tag{3.307}$$

is not accelerated, i.e.,

$$\frac{d^2 \vec{X}}{dt^2} = 0. \tag{3.308}$$

We can choose $\vec{X} = \vec{0}$ such that

$$\vec{x}_1 = \left[\frac{m_2}{m} + \frac{\mu \delta m}{2m^2} \left(\frac{v^2}{c^2} - \frac{m}{r} \right) \right] \vec{x} \tag{3.309}$$

$$\vec{x}_2 = \left[-\frac{m_1}{m} + \frac{\mu \delta m}{2m^2} \left(\frac{v^2}{c^2} - \frac{m}{r} \right) \right] \vec{x} \tag{3.310}$$

where $\vec{x} = \vec{x}_1 - \vec{x}_2, \vec{v} = \vec{v}_1 - \vec{v}_2, m = m_1 + m_2, \delta m = m_1 - m_2, \mu = \frac{m_1 m_2}{m}$ (reduced mass).

For the relative motion we obtain from Eqs. (3.301), (3.302) with (3.309), (3.310) after dividing by μ:

$$\mathcal{L}_{\text{rel}} = \mathcal{L}_0 + \mathcal{L}_1 \tag{3.311}$$

with the Newtonian part $\mathcal{L}_0$ and the post-Newtonian perturbation $\mathcal{L}_1$:

$$\mathcal{L}_0 = \frac{1}{2} \vec{v}^2 + \frac{Gm}{r} \tag{3.312}$$

$$\mathcal{L}_1 = \frac{3}{8} \left(1 - \frac{3\mu}{m} \right) \frac{\vec{v}^4}{c^2} + \frac{Gm}{2rc^2} \left(3\vec{v}^2 + \frac{\mu}{m} \vec{v}^2 + \frac{\mu}{m} \left(\frac{\vec{v}\vec{x}}{r} \right)^2 \right) - \frac{G^2 m^2}{2r^2 c^2}. \tag{3.313}$$

The corresponding Euler–Lagrange equation is ($c = 1$)

$$\dot{\vec{v}} = -\frac{Gm}{r^3} \vec{x} \left(1 - \frac{Gm}{r} \left(4 + \frac{2\mu}{m} \right) + \left(1 + \frac{3\mu}{m} \right) \vec{v}^2 - \left(\frac{3\mu}{2m} \right) \left(\frac{\vec{v}\vec{x}}{r} \right)^2 \right)$$

$$+ \frac{Gm}{r^3} \vec{v}(\vec{v}\vec{x}) \left(4 - \frac{2\mu}{m} \right). \tag{3.314}$$

Fig. 3.5 Elliptical Kepler orbit

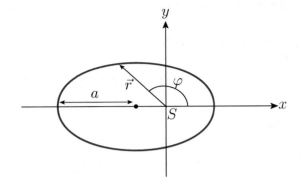

Fig. 3.6 Variable u takes the role of the angle as measured from the center of the ellipse

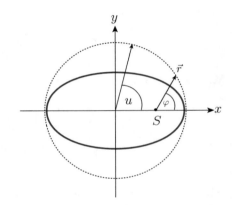

Consider the Kepler problem with motion in the plane $z = 0$ and the periastron lying on the x-axis. Without 1PN corrections one has an elliptic orbit (c.f. Fig. 3.5) where e is the eccentricity and (see Eq. 3.139)

$$r = \frac{a(1 - e^2)}{1 + e \cos \varphi}. \tag{3.315}$$

One finds

$$r = a(1 - e \cos u) \quad \text{where} \quad u - e \sin u = \omega_0 t \tag{3.316}$$

(*Kepler's equation*). Here $\omega_0 = \frac{2\pi}{T}$ with orbital period T ($t = 0$ for passage at periastron) and u is the so-called *eccentric anomaly* ($\omega_0 t$ is the *mean anomaly*). We define $f = \varphi$, the *true anomaly*, such that (c.f. Fig. 3.6)

$$\cos f = \cos \varphi = \frac{\cos u - e}{1 - e \cos u}. \tag{3.317}$$

The invariance under time translations and space rotations of the Lagrangian $\mathcal{L}_{\text{rel}}$ implies four constants of motion—the reduced energy ε and the three components of the reduced angular momentum $\vec{j}$:

$$\varepsilon = \frac{E}{\mu} = \vec{v} \cdot \frac{\partial \mathcal{L}_{\text{rel}}}{\partial \vec{v}} - \mathcal{L}_{\text{rel}}$$

$$= \frac{1}{2}\vec{v}^2 - \frac{Gm}{r} + \frac{3}{8}\left(1 - \frac{3\mu}{m}\right)\frac{\vec{v}^4}{c^2} + \frac{Gm}{2rc^2}\left[\left(3 + \frac{\mu}{m}\right)\vec{v}^2 + \frac{\mu}{m}\left(\frac{\vec{v}\vec{x}}{r}\right)^2\right]$$

$$- \frac{Gm}{r}\bigg], \tag{3.318}$$

$$\vec{j} = \frac{\vec{J}}{\mu} = \vec{x} \wedge \frac{\partial \mathcal{L}_{\text{rel}}}{\partial \vec{v}} = (\vec{x} \wedge \vec{v})\left[1 + \frac{1}{2}\left(1 - \frac{3\mu}{m}\right)\frac{\vec{v}^2}{c^2} + \left(3 + \frac{\mu}{m}\right)\frac{Gm}{rc^2}\right]. \tag{3.319}$$

After some lengthy calculations (including an integration over time), one can get an equation which is analogous to Kepler's equation but with different coefficients:

$$\frac{2\pi}{T_b}t = u - e_t \sin u, \tag{3.320}$$

$$r = a_r(1 - e_r \cos u), \tag{3.321}$$

where

$$a_r = -\frac{Gm}{2\varepsilon}\left[1 - \left(\frac{\mu}{m} - 7\right)\frac{\varepsilon}{2c^2}\right] \tag{3.322}$$

$$e_r^2 = 1 + \frac{2\varepsilon}{G^2m^2}\left[1 + \left(5\frac{\mu}{m} - 15\right)\frac{\varepsilon}{2c^2}\right]\left[j^2 + \left(\frac{\mu}{m} - 6\right)\frac{G^2m^2}{c^2}\right] \tag{3.323}$$

$$e_t^2 = 1 + \frac{2\varepsilon}{G^2m^2}\left[1 + \left(17 - 7\frac{\mu}{m}\right)\frac{\varepsilon}{2c^2}\right]\left[j^2 + \left(2 - \frac{2\mu}{m}\right)\frac{G^2m^2}{c^2}\right] \tag{3.324}$$

$$\frac{2\pi}{T_b} = \frac{(-2\varepsilon)^{3/2}}{Gm}\left[1 - \left(\frac{\mu}{m} - 15\right)\frac{\varepsilon}{4c^2}\right], \tag{3.325}$$

with $j = |\vec{j}|$. Notice that, in a bound orbit, ε is negative. The eccentricity e of the Keplerian orbit is now split into a "radial eccentricity" e_r and a "time eccentricity" e_t. The Newtonian limit is found by considering $c \to \infty$:

$$a_r\big|_{c\to\infty} = \frac{Gm\mu}{-2E} \tag{3.326}$$

$$e_r^2\big|_{c\to\infty} = e_t^2\big|_{c\to\infty} = 1 + \frac{2EL^2}{G^2m^2\mu^3}, \tag{3.327}$$

where L simply denotes the Newtonian limit of $|\vec{J}|$, which is nothing but the classical angular momentum $L \equiv \mu|\vec{r} \wedge \vec{v}|$.

One finds for the true anomaly

$$\varphi(u) = \frac{\cos u - e_\theta}{1 - e_\theta \cos u} \quad \text{or} \quad \varphi(u) = A_{e_\theta}(u) = 2 \arctan\left[\left(\frac{1 + e_\theta}{1 - e_\theta}\right)^{1/2} \tan\left(\frac{u}{2}\right)\right],$$

(3.328)

where $e_\theta = e$ for Kepler orbits. This yields

$$\varphi(u) = \omega_0 + (1 + k)A_{e_\theta}(u) \quad \text{with } k = \frac{3Gm}{c^2 a(1 - e^2)}$$

(3.329)

and $\omega_0^2 = (2\pi/T)^2 = Gm/a^3$. The quantity e_θ is called "angular eccentricity" and it satisfies

$$e_\theta^2 = 1 + \frac{2\varepsilon}{G^2 m^2}\left[1 + \left(\frac{\mu}{m} - 15\right)\frac{\varepsilon}{2c^2}\right]\left[j^2 - \frac{6G^2 m^2}{c^2}\right].$$

(3.330)

For $c \to \infty$, we have as well $e_\theta \to e$.

One can also find the perihelion precession (or periastron precession) per orbit, which is given as follows

$$\boxed{\delta\varphi = \frac{6\pi\, G(m_1 + m_2)}{c^2 a(1 - e^2)}.}$$

(3.331)

Note that here a is the semi-major axis ($a(1 - e^2) = p$). In (3.331) $m = m_1 + m_2$ is the sum of the two masses. In view of the nonlinearities that are involved in the description of the system, this simple result is remarkable.

For the binary pulsar PSR 1913+16 (see Sect. 3.4.6), the measured periastron shift is

$$\dot{\omega}_{\text{obs.}} = 4.226607 \pm 0.00007 \text{ deg yr}^{-1}.$$

(3.332)

The GR prediction which follows from Eq. (3.331) and the known orbital element (given a period of about 7.75 h, i.e. $\dot{\omega} \sim \delta\varphi \times 1130 \text{ deg yr}^{-1}$) is

$$\dot{\omega}_{\text{GR}} = 2.11\left(\frac{m_1 + m_2}{M_\odot}\right)^{2/3} \text{ deg yr}^{-1},$$

(3.333)

where we used

$$a = \left[\frac{G}{\omega_0^2}(m_1 + m_2)\right]^{1/3}$$

(3.334)

Fig. 3.7 Sketch of spatial arrangement of post-Keplerian orbit relative to observer

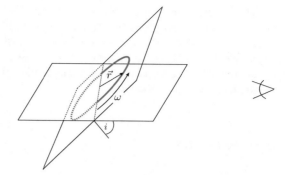

due to Kepler's third law. If we set $\dot{\omega}_{\text{obs.}} = \dot{\omega}_{\text{GR}}$, it follows $m_1 + m_2 = 2.83 M_\odot$.

In order to characterize the pulsar system, several parameters are relevant:

- The parameters which characterize the pulsar itself: right ascension α, declination δ, proper motion, the initial pulse time ϕ_0, the pulse frequency ν, and the spindown parameter.
- The five Keplerian parameters (c.f. Fig. 3.7): T_b (period), t_0 (time of passage at periastron), $x = a_1 \sin i$ (projected semi-major axis, i denotes the inclination angle of the orbital plane with respect to the observer and $a_1 = \frac{m_2 a}{m}$), e (eccentricity), ω (angular position of periastron as measured from the ascending node).
- There are several independent measurable post-Keplerian parameters, we state the most important ones:

  – $\dot{\omega}$ (periastron shift),
  – Δ_E (Einstein time delay) which is related to the transformation from the pulsar proper time to the coordinate time of the pulsar-companion barycenter system. One finds

$$\Delta_E = \gamma \sin u, \tag{3.335}$$

$$\text{where } \gamma = \left(\frac{T_b}{2\pi}\right)^{1/3} e \, \frac{G^{2/3}}{c^2} \frac{m_2(m_1 + 2m_2)}{(m_1 + m_2)^{4/3}}$$

$$= 2.93696 \text{ ms} \cdot \left(\frac{m_2}{M_\odot}\right)\left(\frac{m_1 + 2m_2}{M_\odot}\right)\left(\frac{m_1 + m_2}{M_\odot}\right)^{-4/3} \tag{3.336}$$

  and m_2 is the mass of the companion and m_1 is the mass of the emitting pulsar.
  – $\dot{T}_b$ (see Eq. (3.207)) is due to the emission of gravitational waves. It depends on the masses m_1 and m_2.
  – $r = \frac{Gm_2}{c^3}$ (range of Shapiro time delay) which corresponds to $\frac{a}{c}$ in Eq. (2.87).
  – $s = \sin i = x \, c \, G^{-1/3} \left(\frac{T_b}{2\pi}\right)^{-2/3} m^{2/3} m_2^{-1}$, (shape of Shapiro time delay).
  – $\delta_\theta = \frac{e_\theta - e_t}{e_t} = \frac{G}{c^2 am}\left(\frac{7}{2}m_1^2 + 6m_1 m_2 + 2m_2^2\right)$, where we used Eqs. (3.330) and (3.324).

Seven parameters are needed to fully specify the dynamics of the two-body system (up to an uninteresting rotation about the line of sight). Therefore, the measurement of any two post-Keplerian parameters (besides the five Keplerian parameters) allows to predict the remaining ones. These latter parameters thus constitute a consistency check for GR. For a recent test of GR in a double pulsar see [14], where all the above parameters are nicely discussed.

3.7 Summary

Gravitational waves (GW), predicted 1916 by Einstein, and first detected by the LIGO observatory in 2015, are disturbances in the curvature of spacetime, generated by accelerated masses, that propagate as waves outward from their source at the speed of light. GW are solutions of the linearized Einstein's equations. Similarly to electromagnetic waves GW have also two polarizations transverse to the direction of propagation. Due to the tensorial nature of the equations the two orthogonal states are rotated by 45° rather than 90° as for electromagnetic waves. GW transport energy as gravitational radiation, the rate at which it is emitted from a system due to the change of the mass quadrupole moment is given by Eq. (3.107) derived in 1916 by Einstein. With this formula one can compute the energy loss and thus the shrinking of the orbit of binary neutron star systems. The first such system was discovered by Hulse and Taylor, which nicely confirmed Einstein's quadrupole formula. In the second part of this chapter we discussed the post-Newtonian approximation method, which is very important for computing more accurate gravitational wave forms for the inspiral phase of two coalescing compact objects (black holes, neutron stars or white dwarfs) and for dealing with the observations of binary pulsar systems.

3.8 Problems

Problem 3.1 (*Particles in the field of a gravitational wave*)
Show that the curves $r = r(\varphi)$ described by (Eq. 3.46)

$$r^2(\varphi) = R^2 \begin{cases} 1 - 2h\cos(2\varphi)\cos(\omega t) \\ 1 - 2h\sin(2\varphi)\cos(\omega t) \end{cases} \tag{3.337}$$

for $h \ll 1$ are ellipses. How is the eccentricity e related to h?
Start from the following parametric form of an ellipse

$$r(\varphi) = \frac{b}{\sqrt{1 - e^2\cos^2\varphi}}, \tag{3.338}$$

and assume $e^2 \ll 1$.

Problem 3.2 (*Non-monochromatic gravitational waves*)
Consider monochromatic gravitational waves $h_{\mu\nu}(t, \vec{x}) = h_{\mu\nu}(\vec{x}) \exp[-i\omega t] + \text{c.c.}$.
A general gravitational wave can be a superposition of different frequencies. Start
from the second order Ricci tensor given in Eq. (3.54) to derive the general expression
for the energy-momentum tensor of a gravitational wave imposing Lorenz gauge
$\partial_\mu \gamma^{\mu\nu} = 0$ and tracelessness $\gamma = \eta^{\mu\nu} \gamma_{\mu\nu} = 0$

$$t_{\mu\nu}^{\text{grav}} = \frac{c^4}{32\pi G} \langle \partial_\mu \gamma^{\alpha\beta} \partial_\nu \gamma_{\alpha\beta} \rangle. \tag{3.339}$$

Note that spacetime derivatives ∂_μ in the temporal average can be integrated by parts,
neglecting the boundary terms. The three remaining gauge modes can be used to set
$\gamma^{0i} = 0$, yielding the metric in the transverse traceless gauge $h_{\mu\nu}^{\text{TT}}$. Prove that $h_{00}^{\text{TT}} = 0$
and write down the time–time component of the above energy-momentum tensor in
transverse traceless gauge.

Problem 3.3 (*The quadrupole radiation formula*)
The most general solution for the linearized field equations in the Lorentz gauge
(2.193) is given by

$$\gamma_{\mu\nu}(t, \mathbf{x}) = 4G \int d^3 y \, \frac{1}{|\mathbf{x} - \mathbf{y}|} T_{\mu\nu}(t - |\mathbf{x} - \mathbf{y}|, \mathbf{y}), \tag{3.340}$$

with the retarded time $t_r = t - |\mathbf{x} - \mathbf{y}|$. Show that when the source is isolated, far
away and slowly moving this reduces to the well-known quadrupole radiation formula

$$\gamma_{ij}(t, \vec{x}) = \frac{2G}{r} \frac{d^2 I_{ij}(t_r)}{dt^2}, \tag{3.341}$$

where $I_{ij}(t) = \int d^3 y \, y^i y^j T^{00}(t, \vec{y})$ is the quadrupole moment of the source and
$t_r = t - r$ is the retarded time. Follow this outline:

(a) Put Eq. (3.340) in the frequency domain by the Fourier transforms

$$\tilde{\phi}(\omega, x) = \frac{1}{\sqrt{2\pi}} \int dt e^{-i\omega t} \phi(t, x). \tag{3.342}$$

Change the integration variable from t to $t_r = t - |\vec{x} - \vec{y}|$ by multiplying with
a suitable exponential and note that $dt = dt_r$.
(b) We are interested in the field far away from the isolated source. Approximate
$|\vec{x} - \vec{y}| = r$, where r is just the distance to the source. Also forget about the
time components of the metric perturbation, as they are related to the spatial
components by the Lorenz gauge $\partial_\mu \gamma^{\mu 0} = 0 \Rightarrow \partial_0 \gamma^{00} = -\partial_i \gamma^{i0}$.
(c) Use the product rule $\partial_k(y^j \tilde{T}^{ik}) = y^j \partial_k \tilde{T}^{ik} + \tilde{T}^{ij}$ and the conservation law of
the energy momentum tensor in Fourier space $i\omega \tilde{T}^{0\nu} = -\partial_i \tilde{T}^{i\nu}$ to relate the ij

component to the $i0$ component. Symmetrize the result and use the product rule once more to fully relate the spatial components to the time components.

(d) Transform your equation back to the time domain by applying the inverse Fourier transform

$$\phi(t, x) = \frac{1}{\sqrt{2\pi}} \int d\omega e^{i\omega t} \widetilde{\phi}(\omega, x) . \tag{3.343}$$

Problem 3.4 (*Gravitational Bremsstrahlung*)
The gravitational wave analogue of Bremsstrahlung can be generated by a small mass m scattering off a large mass $M \gg m$ with impact parameter b. Assume that the large mass sits at $(0, 0, 0)$, that $E = 0$ (parabolic orbit) and that the orbit lies in the x-y-plane. Calculate the gravitational wave amplitude at a position on the z-axis.

For the trace reversed metric in Lorenz gauge $\gamma_{\mu\nu} = h_{\mu\nu} - 1/2\, \eta_{\mu\nu} h$ and slowly moving sources the gravitational wave amplitude can be calculated as[7]

$$\gamma_{ij}(t, \vec{x}) = \frac{2G}{rc^4} \frac{d^2 I_{ij}(t_r)}{dt^2} , \tag{3.344}$$

where t_r is the retarded time and $I_{ij}(t) = \int T_{00}(t) y^i y^j \, dy^3$ is the quadrupole tensor. The parabolic orbit of the small mass m is described by the parametric solution

$$r(\varphi) = \frac{2b}{1 + \cos(\varphi)} \qquad \dot{\varphi} = \sqrt{\frac{M}{8b^3}} \left[1 + \cos(\varphi)\right]^2 . \tag{3.345}$$

To find the gravitational wave amplitude, project γ_{ij} to the transverse traceless gauge.

Problem 3.5 (*A spinning rod*)
A thin rod with mass M and length l spins at a frequency ω around one axis.

(a) Let the rod be in the $x - y$ plane, rotating about the $z-$axis (which goes through the center of mass of the rod). Let μ denote the mass per unit length. Calculate the mass quadrupole tensor of the rotating rod.

(b) Calculate the emitted power of gravitational radiation.

(c) Assuming that the rod has a cross-sectional area A, and that the centripetal and electrostatic forces are balanced,

$$e \mid \nabla\phi \mid = rm\omega^2 , \tag{3.346}$$

(where ϕ denotes the electric potential, and m and e are the mass and the charge of the electron, respectively) the charges in the rod will deplete and accumulate

---

[7] See Ref. [15] for derivation and limitations of this formula.

on the endpoints, effectively making the rod an electric quadrupole. Calculate the quadrupole moment Q_E, neglecting the diffuse positive charge in the internal regions of the rod.

(d) Because the charges are accelerating, power is radiated. Express this luminosity $P_E \sim \frac{1}{\varepsilon_0 c^5} \omega^6 Q_E^2$ in terms of the parameters of the system.

(e) If the rod has a density of 10 g/cm³, and rotates at 1 kHz, will electromagnetic or gravitational radiation be the dominant energy loss channel? For this evaluate L_{EM}/L_{GW}.

Problem 3.6 (*Order of magnitude estimates*)

(a) In 2003, a double pulsar, a binary pulsar consisting of two neutron stars (each of mass roughly 1.4 $M_\odot$, with an orbital period of 1.4 h), was discovered. Estimate the timescale for coalescence via emission of gravitational radiation.

(b) Suppose that two supermassive ($10^6\ M_\odot$) black holes merge in a galaxy at a cosmological distance (1 Gpc). Consider a detector built to observe the gravitational waves from such events.

  (i) Estimate the frequency range that the detector would have to operate at.

  (ii) Estimate the strain sensitivity (smallest dimensionless amplitude h) that would be necessary to see mergers at these distances.

  (iii) Estimate the duration of such an event.

Problem 3.7 (*Gravitational waves from a Binary System*)
A binary system with masses m_1 and m_2 is in a circular configuration with radius R. Consider the orbit to be adequately described by Newtonian gravity. We will use this description to compute the leading-order effects of gravitational wave emission. Don't forget that orbits in a problem of this type are most easily described using the "reduced system": A body of mass $\mu = m_1 m_2 / (m_1 + m_2)$ in a circular orbit around a body of mass $M = m_1 + m_2$.

The solution to the trace reversed linearized Einstein equations is given by the quadrupole formula

$$\gamma_{ij}(t, \vec{x}) = \frac{2G}{r} \frac{d^2 I_{ij}}{dt^2}\bigg|_{t_r}. \tag{3.347}$$

The energy loss of such a system is given by

$$P = -\frac{G}{5} \left\langle \frac{d^3 J_{ij}}{dt^3} \frac{d^3 J^{ij}}{dt^3} \right\rangle\bigg|_{t_r}, \tag{3.348}$$

with $J_{ij} = I_{ij} - \frac{1}{3}\delta_{ij}\delta^{kl} I_{kl}$ being the traceless part of the quadrupole moment and the brackets mean averaging over one orbit.

(a) Compute the gravitational wave tensor γ_{ij} as measured by an observer looking down the angular momentum axis of the system.
(b) Compute the rate at which energy is carried away from the system by gravitational waves.
(c) By asserting global conservation of energy in the following form,

$$\frac{\mathrm{d}}{\mathrm{d}t}\left(E_{\text{kinetic}} + E_{\text{potential}} + E_{\text{GW}}\right) = 0, \tag{3.349}$$

derive an equation for $\mathrm{d}R/\mathrm{d}t$, the rate at which the orbital radius shrinks.
(d) Derive the change of the orbital angular frequency ω, caused by the gravitational wave emission. You should find that the masses appear only in the combination $\mu^{3/5}M^{2/5}$, perhaps raised to some power. This combination of masses is known as the "chirp mass", perhaps raised to some power, since it sets the rate at which the frequency chirps.
(e) Integrate the $\mathrm{d}\omega/\mathrm{d}t$ you obtained in part (d), to obtain $\omega(t)$, the time evolution of the binary's orbital frequency. Let T_{coal} (coalescence time), be the time at which the inspiral is over, and the frequency diverges.

Problem 3.8 (*Post-Newtonian Lagrangian for a binary system*)
The gravitational action for a particle a propagating in the metric $g_{\mu\nu}$ is given by

$$S_a = -m_a c \int \mathrm{d}t \left(-g_{\mu\nu}\frac{\mathrm{d}x_a^\mu}{\mathrm{d}t}\frac{\mathrm{d}x_a^\nu}{\mathrm{d}t}\right)^{1/2}. \tag{3.350}$$

In this problem we are considering two masses m_1 and m_2 with separation r. Note that the total action is given by $S = S_1 + S_2$ and that particle 1 propagates in the metric generated by particle 2 and vice versa.

(a) Expand the action to order 1PN.
(b) Expand the energy-momentum tensor

$$T_a^{\mu\nu} = \frac{1}{\sqrt{-g}}\frac{\mathrm{d}t}{\mathrm{d}\tau_a}m_a\frac{\mathrm{d}x_a^\mu}{\mathrm{d}t}\frac{\mathrm{d}x_a^\nu}{\mathrm{d}t}\delta^{(3)}\left(\vec{x} - \vec{x}_a(t)\right) \tag{3.351}$$

to order 1PN. Here τ_a is the proper time of particle a.
(c) Calculate the required metric corrections from the above energy-momentum tensor. Recall that the metric is given order by order as

$$^{(2)}g_{00} = -2\phi \qquad\qquad ^{(2)}g_{ij} = -2\phi\delta_{ij}$$
$$^{(3)}g_{0i} = \xi_i \qquad\qquad ^{(4)}g_{00} = -2\left(\phi^2 + \psi\right) \tag{3.352}$$

where ϕ, ζ_i and ψ are defined as

$$\phi(t, \vec{x}) = -\frac{G}{c^4} \int \frac{d^3x'}{|\vec{x} - \vec{x}'|} \, {}^{(0)}T^{00}(t, \vec{x}')$$ (3.353)

$$\xi_i(t, \vec{x}) = -\frac{4G}{c^4} \int \frac{d^3x'}{|\vec{x} - \vec{x}'|} \, {}^{(1)}T^{0i}(t, \vec{x}')$$ (3.354)

$$\psi(t, \vec{x}) = -\frac{G}{c^4} \int \frac{d^3x'}{|\vec{x} - \vec{x}'|} \left[{}^{(2)}T^{00}(t, \vec{x}') + {}^{(2)}T^{ii}(t, \vec{x}') + \frac{c^4}{4\pi G} \partial_0^2 \phi \right]$$ (3.355)

(d) Plug the metric perturbations into the action and identify the Newtonian and leading-order post-Newtonian Lagrangian.

Problem 3.9 (*Hamiltonian geodesic formulation in PN framework*)
Geodesics are commonly expressed in terms of the Lagrangian,

$$L = \frac{1}{2} g_{\mu\nu} \dot{x}^\mu \dot{x}^\nu,$$ (3.356)

where the dots are derivatives with respect to the affine parameter. An equivalent, yet generally more numerically stable formulation, exists in terms of the Hamiltonian

$$H = \frac{1}{2} g^{\mu\nu} p_\mu p_\nu,$$ (3.357)

with the conjugate momenta

$$p_\mu = \frac{\partial L}{\partial \dot{x}^\mu}.$$ (3.358)

The solutions to Hamilton's equations

$$\dot{x}^\mu = \frac{\partial H}{\partial p_\mu}, \quad \dot{p}_\mu = -\frac{\partial H}{\partial x^\mu},$$ (3.359)

are geodesics.

(a) Write down the Lagrangian and Hamiltonian of the Schwarzschild metric.
(b) Write down Hamilton's equations for a test particle trajectory in the Schwarzschild spacetime.
(c) Expand the equations of motion for both a massive and massless particle to first post-Newtonian order.
(d) Integrate these equations numerically. Produce a plot of a few trajectories. Compare the full solution to the expanded one.

References

1. LIGO Scientific and Virgo Collaborations (B.P. Abbott et al.), Observation of gravitational waves from a binary black hole merger. Phys. Rev. Lett. **116**, 061102 (2016)
2. LIGO Scientific and Virgo Collaborations (B. P. Abbott et al.), GWTC-1: a gravitational-wave transient catalog of compact binary mergers observed by LIGO and virgo during the first and second observing runs. Phys. Rev. X **9**, 031040 (2019)
3. LIGO Scientific and Virgo Collaborations (B.P. Abbott et al.), *GWTC-3: Compact Binary Coalescences Observed by LIGO and Virgo During the Second Part of the Third Observing Run*, arXiv:2111.03606 (2021)
4. The LISA Pathfinder Collaboration, M. Armano et al., Sub-Femto-g free fall for space-based gravitational wave observatories: LISA pathfinder results. Phys. Rev. Lett. **116**, 231101 (2016)
5. The LISA Pathfinder Collaboration, M. Armano et al., Beyond the required LISA free-fall performance: new LISA pathfinder results down to 20 μHz. Phys. Rev. Lett. **120**, 061101 (2018)
6. M. Maggiore, *Gravitational Waves: Volume 1: Theory and Experiments* (Oxford University Press, 2007)
7. P.C. Peters, J. Mathews, Phys. Rev. **131**, 435 (1963)
8. L. De Vittori, P. Jetzer, A. Klein, Phys. Rev. D **86**, 044017 (2012)
9. J. García-Bellido, S. Nesseris, Phys. Dark Univ. **21**, 61 (2018)
10. M. Gröbner, P. Jetzer, M. Haney, S. Tiwari, W. Ishibashi, Class. Quant. Grav. **37**, 067002 (2020)
11. N. Straumann, *General Relativity with Applications to Astrophysics* (Springer, Berlin, 2004)
12. S. Weinberg, *Gravitation and Cosmology* (Wiley, New York, 1972)
13. A. Einstein, L. Infeld, B. Hoffmann, The gravitational equations and the problem of motion. Ann. Math. **39**, 65–100 (1938)
14. M. Kramer, et al., *Strong-field Gravity Tests with the Double Pulsar*, arXiv:2112.06795 (2021)
15. S. Caroll, *Spacetime and Geometry: An Introduction to General Relativity* (Cambridge University Press, 2019)

Chapter 4
Black Holes

In this chapter we discuss different issues on black holes. In particular, we briefly present the Kerr solution for rotating black holes. Moreover, we give an overview of the four laws of black hole dynamics and on Hawking radiation. To describe the Dirac equation in curved spacetimes we need first to introduce the tetrad formalism and the concept of spinors.

4.1 The Kerr Solution

The Kerr solution is one of the most important solutions to Einstein's (vacuum) equations. It describes stationary rotating black holes and was found in 1963 by R. Kerr [1]. Later, it was generalized to the *Kerr–Newman solution* which describes rotating, electrically charged black holes. Here we give an overview on the Kerr and the Kerr–Newman solutions, however, without giving their derivation .

The Kerr solution is axisymmetric and stationary. We use the so-called *Boyer–Lindquist coordinates* (t, r, θ, φ) and the following abbreviations:

$$\Delta = r^2 - 2mr + a^2, \qquad \rho^2 = r^2 + a^2 \cos^2 \theta \, , \tag{4.1}$$

where we set $G = c = 1$. The Kerr metric reads

$$
\begin{aligned}
ds^2 = \frac{1}{\rho^2} &\left[-(\Delta - a^2 \sin^2 \theta) dt^2 + 2a \sin^2 \theta (\Delta - r^2 - a^2) dt d\varphi \right. \\
&\left. + \sin^2 \theta \left((r^2 + a^2)^2 - \Delta a^2 \sin^2 \theta \right) d\varphi^2 \right] + \rho^2 \left[\frac{dr^2}{\Delta} + d\theta^2 \right].
\end{aligned}
\tag{4.2}
$$

© The Author(s), under exclusive license to Springer Nature Switzerland AG 2022
P. Jetzer, *Applications of General Relativity*, UNITEXT for Physics,
https://doi.org/10.1007/978-3-030-95718-6_4

4.1.1   Interpretation of the Parameters a and m

In order to interpret a and m we look at the asymptotic form of the metric (4.2) for large "radial coordinate" r:

$$ds^2 = -\left[1 - \frac{2m}{r} + \mathcal{O}\left(\frac{1}{r^2}\right)\right]dt^2 - \left[\frac{4am}{r}\sin^2\theta + \mathcal{O}\left(\frac{1}{r^2}\right)\right]dtd\varphi$$
$$+ \left[1 + \mathcal{O}\left(\frac{1}{r}\right)\right]\left[dr^2 + r^2(d\theta^2 + \sin^2\theta\, d\varphi^2)\right]. \tag{4.3}$$

The examination is easier by transforming to asymptotically Lorentz coordinates:

$$x = r\sin\theta\cos\varphi, \quad y = r\sin\theta\sin\varphi, \quad z = r\cos\theta. \tag{4.4}$$

This yields the following form of the asymptotic metric:

$$ds^2 = -\left[1 - \frac{2m}{r} + \mathcal{O}\left(\frac{1}{r^2}\right)\right]dt^2 - \left[\frac{4am}{r^3} + \mathcal{O}\left(\frac{1}{r^4}\right)\right](x\, dy - y\, dx)dt$$
$$+ \left[1 + \mathcal{O}\left(\frac{1}{r}\right)\right]\left[dx^2 + dy^2 + dz^2\right], \tag{4.5}$$

where we used $r^2\sin^2\theta\, d\varphi = x\, dy - y\, dx$.

In Sect. 2.6 we computed the metric of the rotating Earth assuming slow rotation. In Eq. (2.162) we found

$$ds^2 = \left[1 - \frac{2GM_E}{c^2r}\right]c^2dt^2 - \left[1 + \frac{2GM_E}{c^2r}\right]d\vec{r}^2 + 2ch_{0i}\, dx^i dt. \tag{4.6}$$

Identifying $\frac{GM_E}{c^2} \to m$, $d\vec{r}^2 = dx^2 + dy^2 + dz^2 = \delta_{ij}dx^i dx^j$ and including an overall sign (since we used another sign convention in Sect. 2.6), this metric should coincide with (4.5).

Using equation (2.157)

$$h_{0i} = \frac{4G}{c^3}\varepsilon_{ikn}\frac{\omega^k x^j}{r^3}\int d^3\tilde{r}\,\tilde{x}^n\rho(\tilde{r})\tilde{x}_j, \tag{4.7}$$

we infer that T^{0i} is proportional to $\rho\frac{v_i}{c}$, where $v_i = \varepsilon_{ikn}\omega^k x^n$ with ω^k being the angular velocity of the rotating body. We define

$$S_k = \varepsilon_{klm}\int d^3x\, x^l T^{m0}, \tag{4.8}$$

which is the intrinsic angular momentum of the rotating body. Therefore,

$$h_{0i} = \frac{4G}{c^3} \varepsilon_{ikm} \frac{x^m S^k}{r^3}.$$ (4.9)

With the sign convention of (4.5), Eqs. (4.5) and (4.6) indeed take the common form

$$ds^2 = -\left[1 - \frac{2m}{r} + \mathcal{O}\left(\frac{1}{r^2}\right) \right] dt^2 + \left[1 + \frac{2m}{r} + \mathcal{O}\left(\frac{1}{r^2}\right) \right] \delta_{ij} \, dx^i dx^j$$
$$- \left[4\varepsilon_{ikl} \frac{S^k x^l}{r^3} + \mathcal{O}\left(\frac{1}{r^4}\right) \right] dt dx^i.$$ (4.10)

Clearly S^k is proportional to the body's mass M and its angular momentum (see Eq. (2.157)):

$$S^k = \frac{aGM}{c^2} \frac{\partial}{\partial z} = a \frac{GM}{c^2} \times \left(\begin{array}{c} \text{unit vector along polar axis} \\ \text{of Boyer–Lindquist coordinates} \end{array} \right).$$ (4.11)

Therefore, m is just the mass, and a can be interpreted as the angular momentum ($0 \le a \le 1$).

4.1.2 Kerr–Newman Solution

The Kerr–Newman solution is the extension of the Kerr solution which also describes electrically charged black holes ($c = G = 1$). We use the three parameters

$$\Delta = r^2 - 2Mr + a^2 + Q^2,$$ (4.12)
$$\rho^2 = r^2 + a^2 \cos^2 \theta,$$ (4.13)
$$\Sigma^2 = (r^2 + a^2)^2 - a^2 \Delta \sin^2 \theta,$$ (4.14)

where M is the total mass, a the intrinsic angular momentum and Q the total charge. The metric coefficients of the Kerr–Newman metric are

$$g_{rr} = \frac{\rho^2}{\Delta}, \qquad g_{\theta\theta} = \rho^2, \qquad g_{\varphi\varphi} = \frac{\Sigma^2}{\rho^2} \sin \theta,$$
$$g_{tt} = -1 + \frac{2Mr - Q^2}{\rho^2}, \qquad g_{t\varphi} = -a \frac{2Mr - Q^2}{\rho^2} \sin^2 \theta,$$

where we assumed $a > 0$ without loss of generality. This metric contains the following special cases:

- $Q = a = 0$: Schwarzschild solution,
- $a = 0$: Reissner–Nordstrøm solution,
- $Q = 0$: Kerr solution.

The electromagnetic field of the Kerr–Newman solution is

$$F = Q\rho^{-4}(r^2 - a^2\cos^2\theta)\,\mathrm{d}r \wedge (\mathrm{d}t - a\sin^2\theta\,\mathrm{d}\varphi)$$
$$+ 2Q\rho^{-4}ar\cos\theta\sin\theta\,\mathrm{d}\theta \wedge \big((r^2 + a^2)\mathrm{d}\varphi - a\mathrm{d}t\big)\,, \tag{4.15}$$

where $\wedge$ denotes the exterior product.

From (4.15) one can deduce the asymptotic expressions for electric and magnetic fields (in r, θ, φ directions)[1]:

$$
\begin{aligned}
E_r &= F_{rt} = \frac{Q}{r^2} + \mathcal{O}\left(\frac{1}{r^3}\right)\,, \\
E_\theta &= \frac{E_\theta}{r} = \frac{F_{\theta t}}{r} = \mathcal{O}\left(\frac{1}{r^4}\right)\,, \\
E_\varphi &= \frac{E_\varphi}{r\sin\theta} = \frac{F_{\varphi t}}{r\sin\theta} = 0\,, \\
B_r &= \frac{F_{\theta\varphi}}{r^2\sin\theta} = \frac{2Qa}{r^3}\cos\theta + \mathcal{O}\left(\frac{1}{r^4}\right)\,, \\
B_\theta &= \frac{F_{\varphi r}}{r\sin\theta} = \frac{Qa}{r^3}\sin\theta + \mathcal{O}\left(\frac{1}{r^4}\right)\,, \\
B_\varphi &= \frac{F_{r\theta}}{r} = 0\,.
\end{aligned}
\tag{4.16}
$$

We see immediately that asymptotically the electric field is a Coulomb field.

4.1.3   Equations of Motion for Test Particles

Let a test particle with electric charge e and rest mass μ move in the external fields of a Kerr–Newman black hole. The equations of motion are

$$\frac{\mathrm{d}^2 x^\alpha}{\mathrm{d}\lambda^2} + \Gamma^\alpha{}_{\beta\gamma}\frac{\mathrm{d}x^\beta}{\mathrm{d}\lambda}\frac{\mathrm{d}x^\gamma}{\mathrm{d}\lambda} = eF^\alpha{}_\beta\frac{\mathrm{d}x^\beta}{\mathrm{d}\lambda}\,. \tag{4.17}$$

The best way to solve this equation turns out to be the Hamiltonian formalism (for details see, e.g., [2, 3]).

---

[1] The components of $\vec{E}$ and $\vec{B}$ are relative to an orthonormal triad (see, e.g., [2]).

We can simplify the analysis by assuming that the metric is that of a Kerr black hole ($Q = 0$) and that the motion is confined to the equatorial plane ($\theta = \pi/2$).[2] In this case, the metric has the following nonvanishing components:

$$g_{tt} = -\left(1 - \frac{2GM}{c^2 r}\right), \tag{4.18}$$

$$g_{t\varphi} = -\frac{a}{r}\frac{2GM}{c^2}, \tag{4.19}$$

$$g_{\varphi\varphi} = \left(r^2 + a^2 + \frac{a^2}{r}\frac{2GM}{c^2}\right). \tag{4.20}$$

Denoting by $K = \frac{E}{\mu c}$ the total energy and by l the angular momentum of the particle, one finds

$$\frac{1}{2}\mu \dot{r}^2 + \mu V_{\text{eff.}} = \text{const.} \tag{4.21}$$

with the effective potential

$$V_{\text{eff.}} = -\frac{GM}{r} + \frac{l^2 - a^2(K^2 - c^2)}{2\mu^2 r^2} - \frac{GM(l - aK)^2}{\mu^2 c^2 r^3}. \tag{4.22}$$

For $a = 0$ this reduces to the Schwarzschild case.

For a black hole which is spinning extremely fast, it can be shown that for a particle which spirals in toward the black hole in an accretion disk from very far away to the innermost circular stable orbit, the fraction $\left(1 - \frac{1}{\sqrt{3}}\right)$ of its rest energy is set free. The innermost stable circular orbit can easily be determined from equation (4.22). Thus a rotating black hole allows a gravitational energy conversion with an efficiency up to $\approx 42.3\%$!

4.2 Tetrad Formalism

4.2.1 Introduction

To derive the effects of gravity on physical systems, we have until now followed the covariance principle: Write the appropriate special relativistic equations that hold in the absence of gravitation, replace $\eta_{\alpha\beta}$ by $g_{\alpha\beta}$, and replace all derivatives with covariant derivatives. The resulting equations will be generally covariant and true in the presence of gravitational fields. However, this method actually works only for objects that behave like tensors under Lorentz transformation and not for spinor

---

[2] Note that the equatorial plane of a rotating black hole is distinguished. For the geodesic equations in the Schwarzschild background any arbitrary plane which includes the origin is equivalent.

fields. We will see that the *tetrad formalism* is useful to incorporate spinor fields (see for more details, e.g., [4]).

The *tetrad formalism* is a different approach to the problem of determining the effects of gravitation on physical systems. As a formalism rather than a theory, it does not make different predictions but does allow the relevant equations to be expressed differently.

In general relativity, coordinate basis methods provide a straightforward procedure to calculate many quantities, for example, ∇_α and $R^\alpha{}_{\mu\beta\nu}$. Sometimes it is of advantage to use an orthonormal basis in tensor calculations. We recall that a coordinate basis $\left\{ \frac{\partial}{\partial x^\mu} \right\}$ is orthonormal only for the trivial case of flat spacetime.

Tetrads are basis vectors which transform the metric onto a Minkowski structure. Due to the principle of equivalence, at every point X, we can erect a set of coordinates ξ_X^α that are locally inertial at X. Note that you can create a *single* inertial coordinate system that is locally inertial everywhere only if the spacetime is flat. The general metric of non-inertial coordinate system is given by:

$$g_{\mu\nu}(x) = \left.\frac{\partial \xi_X^\alpha(x)}{\partial x^\mu}\right|_{x=X} \left.\frac{\partial \xi_X^\beta(x)}{\partial x^\nu}\right|_{x=X} \eta_{\alpha\beta} \equiv V^\alpha{}_\mu(X) V^\beta{}_\nu(X) \eta_{\alpha\beta}, \qquad (4.23)$$

where $V^\alpha{}_\mu(X) := \left.\frac{\partial \xi_X^\alpha(x)}{\partial x^\mu}\right|_{x=X}$. If we want to change our non-inertial coordinates from x^μ to x'^μ, the partial derivatives $V^\alpha{}_\mu$ transform according to the rule

$$V^\alpha{}_\mu \rightarrow V'^\alpha{}_\mu = \frac{\partial x^\nu}{\partial x'^\mu} V^\alpha{}_\nu. \qquad (4.24)$$

We can think of $V^\alpha{}_\mu$ as forming *four* covariant vector fields, rather than a single tensor. This set of four vectors is known as a *tetrad* or *vierbein*.

Given any contravariant vector field $A^\mu(x)$, we can use the tetrad to refer its components at x to the coordinate system ξ_X^α locally inertial at x:

$$*A^\alpha \equiv V^\alpha{}_\mu A^\mu. \qquad (4.25)$$

We contract a contravariant vector A^μ with four covariant vectors $V^\alpha{}_\mu$. This has the effect of replacing the single four-vector A^μ with the four scalars $*A^\alpha$. The same can be done for covariant vector fields and general tensor fields as follows:

$$*A_\alpha \equiv V_\alpha{}^\mu A_\mu, \qquad (4.26a)$$

$$*B^\alpha{}_\beta \equiv V^\alpha{}_\mu V_\beta{}^\nu B^\mu{}_\nu, \qquad (4.26b)$$

where $V_\beta{}^\nu$ is the tetrad $V^\alpha{}_\mu$ with α-index lowered with the Minkowski tensor $\eta_{\alpha\beta}$. The μ-index are raised with the metric tensor $g^{\mu\nu}$ as follows

$$V_\beta{}^\nu = \eta_{\alpha\beta}\, g^{\mu\nu} V^\alpha{}_\mu. \qquad (4.27)$$

According to Eq. (4.23) this is just the inverse of the tetrad:

$$\delta^{\mu}{}_{\nu} = V_{\beta}{}^{\mu} V^{\beta}{}_{\nu} \tag{4.28}$$

(as can easily be verified by using the definition of $V_{\beta}{}^{\mu}$ and Eq. (4.23), so that we get

$$V_{\beta}{}^{\mu} V^{\beta}{}_{\nu} = \eta_{\alpha\beta} g^{\mu\bar{\nu}} V^{\alpha}{}_{\bar{\nu}} V^{\beta}{}_{\nu} = g_{\bar{\nu}\nu} g^{\mu\bar{\nu}} = \delta^{\mu}{}_{\nu}) ,$$

and analogously

$$\delta^{\alpha}{}_{\beta} = V^{\alpha}{}_{\mu} V_{\beta}{}^{\mu} . \tag{4.29}$$

The scalar components of the metric tensor are then:

$$^{*}g_{\alpha\beta} = V_{\alpha}{}^{\mu} V_{\beta}{}^{\nu} g_{\mu\nu} = \eta_{\alpha\beta} . \tag{4.30}$$

This can be seen using the definitions of $V_{\alpha}{}^{\mu}$ and $V^{\mu}{}_{\alpha}$, so that we get

$$V_{\alpha}{}^{\mu} V_{\beta}{}^{\nu} g_{\mu\nu} = \eta_{\alpha\bar{\beta}} g^{\mu\bar{\nu}} V^{\bar{\beta}}{}_{\bar{\nu}} \eta_{\beta\bar{\alpha}} g^{\nu\bar{\mu}} V^{\bar{\alpha}}{}_{\bar{\mu}} g_{\mu\nu} = \eta_{\alpha\bar{\beta}} \eta_{\beta\bar{\alpha}} g^{\mu\bar{\nu}} \delta^{\bar{\mu}}{}_{\mu} V^{\bar{\beta}}{}_{\bar{\nu}} V^{\bar{\alpha}}{}_{\bar{\mu}}$$
$$= \eta_{\alpha\bar{\beta}} V_{\beta}{}^{\bar{\nu}} V^{\bar{\beta}}{}_{\bar{\nu}} = \eta_{\alpha\bar{\beta}} \delta^{\bar{\beta}}{}_{\beta} = \eta_{\alpha\beta} .$$

Note that one could also start defining the tetrad from Eq. (4.30). We have now seen how to make any tensor field into a set of scalars. As mentioned before the tetrad formalism will be useful to incorporate spinor fields, like the Dirac's electron spin 1/2 field.

The equivalence principle requires that special relativity applies to locally inertial frames and that it makes no difference which locally inertial frame we chose. Thus, when going from one local inertial frame at a given point to another at the same point, the fields transform with respect to a Lorentz transformation $\Lambda^{\alpha}{}_{\beta}(x)$ as follows

$$^{*}A^{\alpha}(x) \rightarrow \Lambda^{\alpha}{}_{\beta}(x) \,^{*}A^{\beta}(x) \tag{4.31}$$

$$^{*}T_{\alpha\beta}(x) \rightarrow \Lambda^{\gamma}{}_{\alpha}(x) \Lambda^{\delta}{}_{\beta}(x) \,^{*}T_{\gamma\delta}(x) , \tag{4.32}$$

where

$$\eta_{\alpha\beta} \Lambda^{\alpha}{}_{\gamma}(x) \Lambda^{\beta}{}_{\delta}(x) = \eta_{\gamma\delta} \tag{4.33}$$

satisfies the requirement for the Lorentz transformation. Note that the Lorentz transformation depends on the position and therefore is not a constant.
The tetrad changes according to the same rule of $^{*}A^{\alpha}$:

$$V^{\alpha}{}_{\mu}(x) \rightarrow \Lambda^{\alpha}{}_{\beta}(x) V^{\beta}{}_{\mu}(x) . \tag{4.34}$$

Next we define the spin connection $\omega^{a}{}_{b}$ starting from the Levi-Civita connection

$$(\omega^{a}{}_{b})_{\mu} := V^{a}{}_{\lambda} \nabla_{\mu} V_{b}{}^{\lambda} , \tag{4.35}$$

where the covariant derivative $\nabla_\mu V_b{}^\lambda$ is given by (using the definition of the covariant derivative, see Eq. (1.1)),

$$
\nabla_\mu V_b{}^\lambda = \nabla_{\frac{\partial}{\partial x^\mu}}\left(\frac{\partial}{\partial x_\lambda}\xi_b\right) = \Gamma^\lambda_{\mu\sigma}\frac{\partial}{\partial x_\sigma}\xi_b + \frac{\partial}{\partial x_\lambda}\left(\frac{\partial}{\partial x^\mu}\xi_b\right)
$$
$$
= \Gamma^\lambda_{\mu\sigma}V_b{}^\sigma + \frac{\partial}{\partial x^\mu}V_b{}^\lambda = \partial_\mu V_b{}^\lambda + \Gamma^\lambda_{\mu\sigma}V_b{}^\sigma , \tag{4.36}
$$

and therefore

$$
(\omega^a{}_b)_\mu = V^a{}_\lambda \nabla_\mu V_b{}^\lambda = V^a{}_\lambda\left(V_b{}^\sigma \Gamma^\lambda_{\mu\sigma} + \partial_\mu V_b{}^\lambda\right) = V_b{}^\sigma V^a{}_\lambda \Gamma^\lambda_{\mu\sigma} - V_b{}^\lambda \partial_\mu V^a{}_\lambda
$$
$$
= V_b{}^\sigma\left(V^a{}_\lambda \Gamma^\lambda_{\mu\sigma} - \partial_\mu V^a{}_\sigma\right) , \tag{4.37}
$$

where we used the identity $\delta^a{}_b = V^a{}_\lambda V_b{}^\lambda$ to compute $V^a{}_\lambda \partial_\mu V_b{}^\lambda = \partial_\mu(V^a{}_\lambda V_b{}^\lambda) - V_b{}^\lambda \partial_\mu V^a{}_\lambda = -V_b{}^\lambda \partial_\mu V^a{}_\lambda$.

Note also that a, b are not tensor indices, and they do not transform as such under Lorentz transformations. It follows also that

$$
\nabla_\mu V_b{}^\lambda = (\omega^c{}_b)_\mu V_c{}^\lambda . \tag{4.38}
$$

This relation can be checked through the definition of the 1-form $(\omega^\mu{}_\nu)_a$ as follows

$$
(\omega^a{}_b)_\mu = V^a{}_\lambda \nabla_\mu V_b{}^\lambda = V^a{}_\lambda (\omega^c{}_b)_\mu V_c{}^\lambda = \delta^a{}_c (\omega^c{}_b)_\mu = (\omega^a{}_b)_\mu . \tag{4.39}
$$

Sometimes the tetrad $V_\alpha{}^\mu$ is also denoted with $e_\alpha{}^\mu$ and $V^\alpha{}_\mu$ with $e^\alpha{}_\mu$ (or $e^a{}_\mu$ when α, β are replaced by a, b). Finally note that μ, ν are raised or lowered with $g_{\mu\nu}$ and α, β (a,b) with $\eta_{\alpha\beta}$ (η_{ab}).

4.2.2 Some Differential Geometry

Here we discuss some concepts of differential geometry, for which we follow the notation given in [2].

Let $\mathcal{M}$ be a differentiable manifold with a Levi-Civita connection ∇, $(e_1, \ldots, e_n)$ be a basis of C^∞ vector fields on an open subset U and $(\theta^1, \ldots, \theta^n)$ be the corresponding dual basis of 1-forms, i.e., $\theta^i(e_j) = \delta^i_j$. We use Latin letters to make clear that we work in the e_i frame.

We define the *connection forms* $\omega^i_j \in \Lambda^1(U)$ (space of the completely antisymmetric -p-multi-linear forms) by

$$
\nabla_X e_j = \omega^i_j(X)e_i , \tag{4.40}
$$

$(X \in C^\infty$ is a vector field). We know from the definition of the Christoffel symbols (in the basis e_i) that

$$\nabla_{e_k} e_j = \Gamma^i_{kj} e_i = \omega^i{}_j(e_k) e_i \,, \tag{4.41}$$

thus

$$\omega^i{}_j = \Gamma^i_{kj} \theta^k \,. \tag{4.42}$$

Since ∇_X commutes with contractions, one can show that

$$\nabla_X \theta^i = -\omega^i{}_j(X) \theta^j \,. \tag{4.43}$$

We define the *torsion forms* $\Theta^i \in \Lambda^2(U)$ and *curvature forms* $\Omega^i{}_j \in \Lambda^2(U)$ by

$$T(X, Y) = \Theta^i(X, Y) e_i \,, \tag{4.44a}$$
$$R(X, Y) e_j = \Omega^i{}_j(X, Y) e_i \,. \tag{4.44b}$$

We can expand $\Omega^i{}_j$ as follows

$$\Omega^i{}_j = \frac{1}{2} R^i{}_{jkl} \theta^k \otimes \theta^l \,. \tag{4.45}$$

The torsion forms and curvature forms satisfy the *Cartan structure equations*:

$$dg_{ij} = \omega_{ij} + \omega_{ji} \,, \tag{4.46a}$$
$$\Theta^i = d\theta^i + \omega^i{}_j \wedge \theta^j \,, \tag{4.46b}$$
$$\Omega^i{}_j = d\omega^i{}_j + \omega^i{}_k \wedge \omega^k{}_j \,, \tag{4.46c}$$

where $\omega_{ij} = g_{ik} \omega^k{}_j$ and $dg_{ij} = e_k(g_{ij}) \theta^k = \partial_k(g_{ij}) dx^k$ is the exterior derivative of the components of the metric in the basis e_i.

Let $\mathcal{M}$ now be a 4-dimensional Lorentzian manifold with orthonormal basis $(e_0, \ldots, e_3)$ and dual basis $(\theta^0, \ldots, \theta^3)$, i.e., such that $g(e_i, e_j) = \eta_{ij}$. In these coordinates we can write the metric as $g = \eta_{ij} \theta^i \otimes \theta^j$. In this case the Cartan structure equations simplify to

$$\omega_{ij} = -\omega_{ji} \,, \tag{4.47a}$$
$$d\theta^i = -\omega^i{}_j \wedge \theta^j \,, \tag{4.47b}$$
$$\Omega^i{}_j = d\omega^i{}_j + \omega^i{}_k \wedge \omega^k{}_j \,. \tag{4.47c}$$

Note also that $\Omega_{ij} = \eta_{ik} \Omega^k{}_j = -\Omega_{ji}$.

4.2.3   The Schwarzschild Metric Revisited

The Cartan structure equations allow us to easily calculate the Riemann tensor and its contractions. We will now rederive the Schwarzschild metric using the tetrad formalism.

The static isotropic metric can be written in the form (see (2.37))

$$g = -e^{2a(r)}dt^2 + e^{2b(r)}dr^2 + r^2 d\theta^2 + r^2 \sin^2(\theta) d\phi^2 . \tag{4.48}$$

The metric is asymptotically flat, thus $a(r \to \infty) = b(r \to \infty) = 0$.

4.2.3.1   Orthonormal Coordinates

The Schwarzschild coordinates (t, r, θ, ϕ) are well adapted to the symmetries of the spacetime, but they are not orthonormal, as $g(\partial_i, \partial_j) = g_{ij} \neq \eta_{ij}$. We thus construct an orthonormal frame (e_0, e_1, e_2, e_3) at every point in the spacetime (i.e., a tetrad field), such that $g(e_i, e_j) = \eta_{ij}$. By observation of Eq.(4.48) we find

$$e_0 = e^{-a} \partial_t , \quad e_2 = r^{-1} \partial_\theta ,$$
$$e_1 = e^{-b} \partial_r , \quad e_3 = (r \sin \theta)^{-1} \partial_\phi . \tag{4.49}$$

The corresponding co-frame $(\theta^0, \theta^1, \theta^2, \theta^3)$ is found by the condition $\theta^i(e_j) = \delta^i_j$, so

$$\theta^0 = e^a \, dt , \quad \theta^2 = r \, d\theta ,$$
$$\theta^1 = e^b \, dr , \quad \theta^3 = r \sin \theta \, d\phi . \tag{4.50}$$

4.2.3.2   Connection Forms

We now want to solve the Cartan structure equations (4.47), meaning we have to calculate the exterior derivatives $d\theta^i$. We find

$$d\theta^0 = d(e^a dt) = \partial_i(e^a) \, dx^i \wedge dt = \partial_r(e^a) \, dr \wedge dt = a' e^a \, dr \wedge dt ,$$
$$d\theta^1 = 0 ,$$
$$d\theta^2 = dr \wedge d\theta ,$$
$$d\theta^3 = \sin \theta \, dr \wedge d\phi + r \cos \theta \, d\theta \wedge d\phi . \tag{4.51}$$

We want to work in the basis given by the θ^i, so we rewrite these as follows:

$$d\theta^0 = -a'e^{-b}\,\theta^0 \wedge \theta^1\,,$$

$$d\theta^1 = 0\,,$$

$$d\theta^2 = r^{-1}e^{-b}\,\theta^1 \wedge \theta^2\,,$$

$$d\theta^3 = r^{-1}e^{-b}\,\theta^1 \wedge \theta^3 + r^{-1}\cot\theta\,\theta^2 \wedge \theta^3\,. \tag{4.52}$$

From the first and second Cartan equation we now can read of the connection forms. For example, we have

$$a'e^{-b}\,\theta^0 \wedge \theta^1 = \omega^0_{\ 1} \wedge \theta^1 + \omega^0_{\ 2} \wedge \theta^2 + \omega^0_{\ 3} \wedge \theta^3\,. \tag{4.53}$$

From this we can read of that

$$\omega^0_{\ 1} = \omega^1_{\ 0} = -a'e^{-b}\theta^0 + c_1\theta^1\,,$$

$$\omega^0_{\ 2} = \omega^2_{\ 0} = c_2\theta^2\,,$$

$$\omega^0_{\ 3} = \omega^3_{\ 0} = c_3\theta^3\,, \tag{4.54}$$

where we cannot yet determine the constants c_i due to the fact that $\theta^i \wedge \theta^i = 0$. Solving the other equations we finally find

$$\omega^0_{\ 1} = \omega^1_{\ 0} = a'e^{-b}\theta^0\,,$$

$$\omega^0_{\ 2} = \omega^2_{\ 0} = 0\,,$$

$$\omega^0_{\ 3} = \omega^3_{\ 0} = 0\,,$$

$$\omega^1_{\ 2} = -\omega^2_{\ 1} = -r^{-1}e^{-b}\theta^2\,,$$

$$\omega^1_{\ 3} = -\omega^3_{\ 1} = -r^{-1}e^{-b}\theta^3\,,$$

$$\omega^2_{\ 3} = -\omega^3_{\ 2} = -r^{-1}\cot\theta\,\theta^3\,. \tag{4.55}$$

4.2.3.3 Curvature Forms

We are now in a place to calculate the curvature forms from the third Cartan structure equation (4.47c). For example, we find

$$\Omega^0_{\ 1} = d\omega^0_{\ 1} + \omega^0_{\ 2} \wedge \omega^2_{\ 1} + \omega^0_{\ 3} \wedge \omega^3_{\ 1} = d\omega^0_{\ 1} = d(a'e^{-b}\theta^0) = d(a'e^{-b}) \wedge \theta^0 + a'e^{-b}d\theta^0$$

$$= \partial_r(a'e^{-b})dr \wedge \theta^0 - a'^2e^{-2b}\,\theta^0 \wedge \theta^1$$

$$= -e^{-2b}(a'^2 - a'b' + a'')\,\theta^0 \wedge \theta^1\,, \tag{4.56}$$

where we used Eq. (4.50) for the last equality. The other curvature forms follow similarly, and the results are

$$\Omega^0{}_1 = \Omega^1{}_0 = -e^{-2b}(a'^2 - a'b' + a'')\,\theta^0 \wedge \theta^1\,,$$

$$\Omega^0{}_2 = \Omega^2{}_0 = -\frac{a'e^{-2b}}{r}\,\theta^0 \wedge \theta^2\,,$$

$$\Omega^0{}_3 = \Omega^3{}_0 = -\frac{a'e^{-2b}}{r}\,\theta^0 \wedge \theta^3\,,$$

$$\Omega^1{}_2 = -\Omega^2{}_1 = \frac{b'e^{-2b}}{r}\,\theta^1 \wedge \theta^2\,,$$

$$\Omega^1{}_3 = -\Omega^3{}_1 = \frac{b'e^{-2b}}{r}\,\theta^1 \wedge \theta^3\,,$$

$$\Omega^2{}_3 = -\Omega^3{}_2 = \frac{1 - e^{-2b}}{r^2}\,\theta^2 \wedge \theta^3\,. \tag{4.57}$$

4.2.3.4   Ricci Tensor

The Ricci tensor is given by the contraction of the Riemann tensor. Using the curvature forms we find

$$R_{ij} = \theta^l\left(R(e_l, e_j)e_i\right) \overset{(4.44b)}{=} \theta^l\left(\Omega^k{}_i(e_l, e_j)e_k\right) = \Omega^l{}_i(e_l, e_j)\,. \tag{4.58}$$

We can thus easily calculate the components of the Ricci tensor. For example, the 00-component is given by

$$R_{00} = \Omega^1{}_0(e_1, e_0) + \Omega^2{}_0(e_2, e_0) + \Omega^3{}_0(e_3, e_0)\,. \tag{4.59}$$

Note that $\theta^i \wedge \theta^j = \theta^i \otimes \theta^j - \theta^j \otimes \theta^i$ and thus

$$\theta^i \wedge \theta^j(e_k, e_l) = \theta^i \otimes \theta^j(e_k, e_l) - \theta^j \otimes \theta^i(e_k, e_l) = \delta^i_k\delta^j_l - \delta^j_k\delta^i_l\,. \tag{4.60}$$

The 00-component of the Ricci tensor then evaluates to

$$R_{00} = e^{-2b}(a'^2 - a'b' + a'') + \frac{a'e^{-2b}}{r} + \frac{a'e^{-2b}}{r} = e^{-2b}\left(a'^2 - a'b' + a'' + \frac{2a'}{r}\right)\,. \tag{4.61}$$

The other components evaluate similarly to

$$R_{11} = -e^{-2b}\left(a'^2 - a'b' + a'' - \frac{2b'}{r}\right), \tag{4.62a}$$

$$R_{22} = e^{-2b}\left(\frac{b' - a'}{r} - \frac{1}{r^2}\right) + \frac{1}{r^2}, \tag{4.62b}$$

$$R_{33} = R_{22}. \tag{4.62c}$$

The Ricci scalar then is

$$\mathcal{R} = \eta^{ij}R_{ij} = -2e^{-2b}\left(a'^2 - a'b' + a'' + \frac{2(a' - b')}{r} + \frac{1}{r^2}\right) + \frac{2}{r^2}. \tag{4.63}$$

The non-diagonal terms are trivially zero. The components of the Einstein tensor are computed using $G_{ij} = R_{ij} - \frac{1}{2}\eta_{ij}\mathcal{R}$

$$G_{00} = \frac{1}{r^2} + e^{-2b}\left(\frac{2b'}{r} - \frac{1}{r^2}\right), \tag{4.64a}$$

$$G_{11} = -\frac{1}{r^2} + e^{-2b}\left(\frac{2a'}{r} + \frac{1}{r^2}\right), \tag{4.64b}$$

$$G_{22} = e^{-2b}\left(a'^2 - a'b' + a'' + \frac{1}{r}(a' - b')\right), \tag{4.64c}$$

$$G_{33} = G_{22}, \tag{4.64d}$$

$$G_{\mu\nu} = 0 \quad \text{for all other components.} \tag{4.64e}$$

4.2.3.5 Solving the Einstein Field Equations

We now want to solve the vacuum field equations $G_{ij} = 0$. We combine the first two of these equations to find

$$G_{00} + G_{11} = \frac{2e^{-2b}}{r}(a' + b') = 0. \tag{4.65}$$

This is equal to $a' + b' = 0$ or $a + b = \text{const} = 0$. The 00-equation reads

$$G_{00} = \frac{1}{r^2} + e^{-2b}\left(\frac{2b'}{r} - \frac{1}{r^2}\right) = 0, \tag{4.66}$$

which is equivalent to

$$\frac{d}{dr}\left(re^{-2b}\right) = 1 \quad \Rightarrow \quad e^{-2b} = 1 - \frac{2m}{r}. \tag{4.67}$$

m is an integration constant and can be determined by the Newton limit $m = GM/c^2$. The static isotropic metric thus reads

$$g = -\left(1 - \frac{2m}{r}\right)dt^2 + \frac{1}{\left(1 - \frac{2m}{r}\right)}dr^2 + r^2 d\theta^2 + r^2 \sin^2(\theta)d\phi^2 . \qquad (4.68)$$

We see here the advantage of the tetrad formalism: Instead of computing 32 Christoffel symbols, we only had to compute 6 curvature forms to get the same result.

4.3   Spinors in Curved Spacetime

For this section we assume that the reader has some knowledge of group theory, spinors and the Dirac equation. For more details we refer, e.g., to [4, 5].

4.3.1   Representations of the Lorentz Group

Under a general Lorentz transformation a set of quantities ψ_n transform into the new quantities according to

$$\psi'_n = \sum_m \left[D(\Lambda)\right]_{nm} \psi_m , \qquad (4.69)$$

where Λ is a Lorentz transformation (LT) and $D(\Lambda)$ is a *representation* of the Lorentz group. A representation has to fulfill the following property

$$D(\Lambda_1)D(\Lambda_2) = D(\Lambda_1\Lambda_2) , \qquad (4.70)$$

which means that the Lorentz transformation Λ_1 followed by the Lorentz transformation Λ_2 gives the same result as the Lorentz transformation $\Lambda_1\Lambda_2$. Equation (4.70) is understood as a matrix multiplication. For instance, if ψ is a contravariant vector V^β, then $D(\Lambda)$ is

$$\left[D(\Lambda)\right]^\alpha{}_\beta = \Lambda^\alpha{}_\beta , \qquad (4.71)$$

while if ψ is a covariant tensor $S_{\alpha\beta}$, the corresponding D-matrix is

$$\left[D(\Lambda)\right]_{\gamma\delta}{}^{\alpha\beta} = \Lambda_\gamma{}^\alpha \Lambda_\delta{}^\beta . \qquad (4.72)$$

One can verify that Eqs. (4.71) and (4.72) satisfy the group multiplication rule (4.70).

The most general true representations of the homogeneous Lorentz group are provided by tensor representations such as Eqs. (4.71) and (4.72). Hence, we might expect that all quantities of physical interest should be tensors. However, there are additional representations of the *infinitesimal* Lorentz group and the *spinor representations* that play an important role in quantum field theory. The infinitesimal Lorentz group consists of Lorentz transformations, which are infinitesimally close to the identity, described by

$$\Lambda^\mu{}_\nu = \delta^\mu{}_\nu + \alpha^\mu{}_\nu \,, \tag{4.73}$$

where $|\alpha^\mu{}_\nu| \ll 1$. Furthermore, to first order in α, the condition $\eta_{\mu\nu}\Lambda^\mu{}_\rho\Lambda^\nu{}_\sigma = \eta_{\rho\sigma}$, valid for Lorentz transformations, implies that the matrix is antisymmetric $\alpha_{\mu\nu} = -\alpha_{\nu\mu}$.

We now define the *generators* $T_{\mu\nu}$ of the infinitesimal Lorentz group using the representation of the infinitesimal Lorentz transformation

$$D(\Lambda) = 1 + \frac{1}{2}\alpha_{\mu\nu}T^{\mu\nu} \,, \tag{4.74}$$

where $T_{\mu\nu}$ are a fixed set of antisymmetric matrices, i.e., $T^{\mu\nu} = -T^{\nu\mu}$.

It turns out that the generators $T^{\alpha\beta}$ of the infinitesimal Lorentz group for a contravariant vector have the components

$$(T^{\alpha\beta})^\mu{}_\nu = \eta^{\alpha\mu}\delta^\beta{}_\nu - \eta^{\mu\beta}\delta^\alpha{}_\nu \,. \tag{4.75}$$

For example, one can use this result to compute the transformation of the vector x^μ and recover (with equation (4.71)) $x'^\mu = \Lambda^\mu{}_\nu x^\nu = x^\mu + \frac{1}{2}\alpha_{\rho\sigma}(T^{\rho\sigma})^\mu{}_\nu x^\nu = x^\mu + \alpha^\mu{}_\nu x^\nu$. On the other hand the generators of a covariant tensor have components

$$\left[T_{\alpha\beta}\right]_{\gamma\delta}{}^{\varepsilon\xi} = \eta_{\alpha\gamma}\delta^\varepsilon{}_\beta\delta^\xi{}_\delta - \eta_{\beta\gamma}\delta^\varepsilon{}_\alpha\delta^\xi{}_\delta + \eta_{\alpha\delta}\delta^\xi{}_\beta\delta^\varepsilon{}_\gamma - \eta_{\beta\delta}\delta^\xi{}_\alpha\delta^\varepsilon{}_\gamma \,. \tag{4.76}$$

Furthermore it can be shown that $T_{\alpha\beta}$ satisfies the commutation relations

$$\left[T_{\alpha\beta}, T_{\gamma\delta}\right] = \eta_{\gamma\beta}T_{\alpha\delta} - \eta_{\gamma\alpha}T_{\beta\delta} + \eta_{\delta\beta}T_{\gamma\alpha} - \eta_{\delta\alpha}T_{\gamma\beta} \,. \tag{4.77}$$

One can easily check this inserting the matrices (4.75) and (4.76) in the above relation. Thus the problem of finding the general representation of the infinitesimal homogeneous Lorentz group is reduced to finding all matrices which satisfy Eq. (4.77).

Next we have to consider how a quantity ψ_n transforms under a position-dependent Lorentz transformation or a more general coordinate transformation. Under the transformation $x^\mu \to x'^\mu$ the derivative changes according to

$$\frac{\partial}{\partial x^\mu} \to \frac{\partial}{\partial x'^\mu} = \frac{\partial x^\nu}{\partial x'^\mu}\frac{\partial}{\partial x^\nu} \,. \tag{4.78}$$

Remember that ψ_n transforms according to Eq. (4.69), but now Λ depends on the position and thus we need to extend the definition of covariant derivative to a general representation D. Note that in the case of tensor fields, this is just the covariant derivative. Here we give the final result.[3]

The covariant derivative of a field ψ transforming in a representation of the Lorentz group with generators $T^{\mu\nu}$ is, in an orthonormal basis,

$$\nabla_\mu \psi = \partial_\mu \psi + \frac{1}{2}(\omega_{\nu\rho})_\mu T^{\nu\rho}\psi \,, \tag{4.79}$$

where $\partial_\mu \psi = V^b{}_\mu \partial_b \psi$ and $(\omega_{\nu\rho})_\mu$ is given in Eq. (4.35). Notice that in [4] a different notation is used, with respect to the one we are using: Our $T^{\mu\nu}$ corresponds to $\sigma^{\mu\nu}$ in [4] and thus

$$\widetilde{\Gamma}_\mu \equiv \frac{1}{2}\sigma^{\nu\rho}V_\nu{}^\delta V_{\rho\delta;\mu} = \frac{1}{2}T^{\nu\rho}(\omega_{\nu\rho})_\mu \,. \tag{4.80}$$

Note that $\widetilde{\Gamma}_\mu$ is a matrix.[4] One can introduce a coordinate-scalar Lorentz vector derivative with the tetrad $V_\alpha{}^\mu$ as follows

$$
\begin{aligned}
\mathscr{D}_\alpha \equiv V_\alpha{}^\mu \mathcal{D}_\mu = V_\alpha{}^\mu \Big[\frac{\partial}{\partial x^\mu} + \widetilde{\Gamma}_\mu\Big] &= V_\alpha{}^\mu \frac{\partial}{\partial x^\mu} + \frac{1}{2}\sigma^{\nu\rho}V_\nu{}^\delta V_\alpha{}^\mu V_{\rho\delta;\mu} \\
&= V_\alpha{}^\mu \frac{\partial}{\partial x^\mu} + \frac{1}{2}T^{\nu\rho}V_\alpha{}^\mu(\omega_{\nu\rho})_\mu \,.
\end{aligned}
\tag{4.81}
$$

The expression for $\widetilde{\Gamma}_\mu$ is found by considering infinitesimal Lorentz transformations. The effect of gravity is such that the equations, which hold in special relativity, are modified by replacing $\frac{\partial}{\partial x^\mu}$ with $\mathscr{D}_\alpha$, where $\mathcal{D}_\mu$ is the covariant derivative for fermionic fields.

4.4 Dirac Gamma Matrices

The Dirac gamma matrices are a set of square matrices $\{\Gamma^\mu\}$ which obey the anti-commutation relation

$$\{\Gamma^\mu, \Gamma^\nu\} = \Gamma^\mu\Gamma^\nu + \Gamma^\nu\Gamma^\mu = 2\eta^{\mu\nu}I_4 \,, \tag{4.82}$$

where $\{\ \}$ denotes the anticommutator and I_4 the 4×4 identity matrix. Covariant gamma matrices are defined by

$$\Gamma_\mu = \eta_{\mu\nu}\Gamma^\nu = (\Gamma^0, -\Gamma^1, -\Gamma^2, -\Gamma^3) \,. \tag{4.83}$$

In 4-dimensional spacetime, the smallest representation of the gamma matrices is given by the Dirac representation in which Γ^μ are the 4×4 matrices

---

[3] See Weinberg's book [4], pp. 368–370 for the explicit derivation.

[4] Not to be confused with the Christoffel symbols nor with the Dirac matrices.

$$\Gamma^0 = \begin{bmatrix} I_2 & 0 \\ 0 & -I_2 \end{bmatrix}, \Gamma^1 = \begin{bmatrix} 0 & \sigma_x \\ -\sigma_x & 0 \end{bmatrix}, \Gamma^2 = \begin{bmatrix} 0 & \sigma_y \\ -\sigma_y & 0 \end{bmatrix}, \Gamma^3 = \begin{bmatrix} 0 & \sigma_z \\ -\sigma_z & 0 \end{bmatrix}, \quad (4.84)$$

where σ_x, σ_y, σ_z are the 2×2 Pauli matrices and I_2 is the 2×2 identity matrix. The Dirac representation is used to describe spin $1/2$ particles.

Furthermore, the matrices

$$T^{\mu\nu} = -\frac{1}{4}[\Gamma^\mu, \Gamma^\nu] \qquad (4.85)$$

form a representation of the Lorentz algebra, as given by Eq. (4.77).

4.4.1 Dirac Equation in Flat Space

The Dirac equation for a free particle of mass m reads

$$\left(i\hbar\Gamma^\mu\partial_\mu - mc\right)\psi = 0 \quad \Leftrightarrow \quad (i\hbar\slashed{\partial} - mc)\psi = 0, \qquad (4.86)$$

where $\psi = (\psi_1, \psi_2)$ is a bispinor consisting of two spinors, ψ_1 and ψ_2, and $\slashed{\partial} = \Gamma^\mu\partial_\mu$ is the Feynman slash notation.

The Dirac equation for a particle in an external electromagnetic field, described by the four potential $A_\mu(x)$, is obtained by replacing ∂_μ with $\partial_\mu + i\frac{e}{c\hbar}A_\mu$. In natural units ($\hbar = c = 1$) we obtain

$$(i\slashed{\partial} - e\slashed{A} - m)\psi = 0. \qquad (4.87)$$

This equation describes, for instance, the hydrogen atom.

4.5 Dirac Equation in Curved Spacetime

Since we are dealing with spinors in curved spacetime we replace ∂_μ by $\mathscr{D}_a$. The tetrad defines a local rest frame, allowing the constant Dirac matrices to act at each spacetime point. We thus get

$$(i\hbar\Gamma^a V_a{}^\mu \mathscr{D}_\mu - mc)\psi = 0, \qquad (4.88)$$

with $V_a{}^\mu\mathscr{D}_\mu = \mathscr{D}_a$. If there is an external electromagnetic field, we have the extra term $-\frac{e}{c}\Gamma^\mu A_\mu$ as in equation (4.87)

$$(i\hbar\Gamma^a V_a{}^\mu \mathscr{D}_\mu - \frac{e}{c}\Gamma^\mu A_\mu - mc)\psi = 0, \qquad (4.89)$$

where A_μ is the four potential.

4.5.1   Dirac Equation in Kerr–Newman Geometry

Here we study the Dirac equation around a charged, rotating black hole, whose metric is given by the Kerr–Newman one. We will not give here the full derivation, but just list the main results (for a derivation we refer to [6, 7]). Let P^B and Q^B be two-component spinors representing the wave functions of particles and antiparticles, respectively, then

$$\psi = \begin{bmatrix} P^B \\ Q^B \end{bmatrix} . \tag{4.90}$$

For the Kerr–Newman black hole we consider the Boyer–Lindquist coordinates given in terms of the parameters defined in (4.12)–(4.14).[5]

The Dirac equation in natural units takes the following form

$$\sqrt{2}\left(\nabla_{BB'} + \mathrm{i.e.} A_{BB'}\right)P^B + i\mu_e Q^*{}_{B'} = 0 , \tag{4.91}$$

$$\sqrt{2}\left(\nabla_{BB'} - \mathrm{i.e.} A_{BB'}\right)Q^B + i\mu_e P^*{}_{B'} = 0 , \tag{4.92}$$

where μ_e is the electron (or particle) mass and $A_{BB'}$ is the electromagnetic potential, which in the second equation enters with a minus sign to preserve gauge invariance, due to the fact that the spinors in the two equations are related by complex conjugation. Here the asterisk denotes the complex conjugate,[6] and ∇ the covariant derivative. The indices $A, B, A', B' = \{0, 1\}$ denote the spinor components.

To lower and raise indices of a spinor[7] we use the two-dimensional Levi-Civita symbol ε_{AB} and ε^{AB}, respectively (where $\varepsilon_{00} = \varepsilon_{11} = 0$ and $\varepsilon_{01} = -\varepsilon_{10} = 1$), thus

$$P^A = \varepsilon^{AB} P_B , \quad P_A = \varepsilon_{AB} P^B . \tag{4.93}$$

We shall assume that the four components of the wave function have a dependence on time t and on the azimuthal angle ϕ given by

$$e^{i(\sigma t + m\phi)} . \tag{4.94}$$

This dependence is that of a wave function with frequency σ (or energy) and axial angular momentum m. Then one can make the following ansatz to solve the Eqs. (4.91) and (4.92)

---

[5] Reference [7] is using a different notation: $\rho \leftrightarrow \Sigma$.

[6] In [6] the over-bar is used instead of the asterisk.

[7] See Chandrasekhar's book [5] p. 534.

$$P^0 = (r - ia\cos\theta)^{-1} e^{i(\sigma t + m\phi)} f_1(r, \theta)\,, \tag{4.95}$$

$$P^1 = e^{i(\sigma t + m\phi)} f_2(r, \theta)\,, \tag{4.96}$$

$$Q^{*1} = e^{i(\sigma t + m\phi)} g_1(r, \theta)\,, \tag{4.97}$$

$$Q^{*0} = -(r + ia\cos\theta)^{-1} e^{i(\sigma t + m\phi)} g_2(r, \theta)\,. \tag{4.98}$$

Let us define the following quantity

$$K := (r^2 + a^2)\sigma + am - eQr\,, \tag{4.99}$$

and make the separation ansatz for f_1, f_2, g_1 and g_2:

$$f_1(r, \theta) = R_{-\frac{1}{2}}(r) S_{-\frac{1}{2}}(\theta)\,, \tag{4.100}$$

$$f_2(r, \theta) = R_{+\frac{1}{2}}(r) S_{+\frac{1}{2}}(\theta)\,, \tag{4.101}$$

$$g_1(r, \theta) = R_{+\frac{1}{2}}(r) S_{-\frac{1}{2}}(\theta)\,, \tag{4.102}$$

$$g_2(r, \theta) = R_{-\frac{1}{2}}(r) S_{+\frac{1}{2}}(\theta)\,, \tag{4.103}$$

where $R_{\pm\frac{1}{2}}(r)$ and $S_{\pm\frac{1}{2}}(\theta)$ are functions of r and θ, respectively, with $R_{+\frac{1}{2}}(r)$ the radial wave function for the particle with spin up. Inserting the ansatz into the Eqs. (4.91) and (4.92) gets us four constants of separation λ_1, λ_2, λ_3, λ_4 which have to satisfy for consistency $\lambda_1 = \lambda_3 = \frac{1}{2}\lambda_2 = \frac{1}{2}\lambda_4 = \lambda$. To write these equations, we use the following definitions

$$\mathscr{D}_n := \partial_r + \frac{iK}{\Delta} + 2n\frac{r - M}{\Delta}\,, \tag{4.104}$$

$$\mathscr{D}_n^\dagger := \partial_r - \frac{iK}{\Delta} + 2n\frac{r - M}{\Delta}\,, \tag{4.105}$$

$$\mathscr{L}_n := \partial_\theta + \tilde{Q} + n\cot\theta\,, \tag{4.106}$$

$$\mathscr{L}_n^\dagger := \partial_\theta - \tilde{Q} + n\operatorname{cosec}\theta\,, \tag{4.107}$$

where $\tilde{Q} = a\sigma\sin\theta + m\operatorname{cosec}\theta$. Note that $\mathscr{D}_n$, $\mathscr{D}_n^\dagger$ are purely radial operators, while $\mathscr{L}_n$, $\mathscr{L}_n^\dagger$ are purely angular operators. Note also that r and M have the same dimension in natural units. Furthermore when these operators are applied to "background" quantities, i.e., quantities independent of t and ϕ, they reduce to ∂_r and ∂_θ, respectively.[8] One gets then the following equations

---

[8] See Chandrasekhar's book [5] p. 383.

$$\mathscr{D}_0 R_{-\frac{1}{2}} = (\lambda + i\mu_e r) R_{+\frac{1}{2}}(r) \,, \tag{4.108}$$

$$\Delta \mathscr{D}_{\frac{1}{2}}^\dagger R_{+\frac{1}{2}} = 2(\lambda - i\mu_e r) R_{-\frac{1}{2}}(r) \,, \tag{4.109}$$

$$\mathscr{L}_{\frac{1}{2}} S_{+\frac{1}{2}} = -\sqrt{2}(\lambda - a\mu_e \cos\theta) S_{-\frac{1}{2}}(\theta) \,, \tag{4.110}$$

$$\mathscr{L}_{\frac{1}{2}}^\dagger S_{-\frac{1}{2}} = \sqrt{2}(\lambda + a\mu_e \cos\theta) S_{+\frac{1}{2}}(\theta) \,. \tag{4.111}$$

We can for instance eliminate $R_{+\frac{1}{2}}(r)$ from the first pair of equations to get an equation for $R_{-\frac{1}{2}}(r)$, as follows

$$\left[\Delta \mathscr{D}_{\frac{1}{2}}^\dagger \mathscr{D}_0 - \frac{i\mu_e \Delta}{\lambda + i\mu_e r} \mathscr{D}_0 - 2(\lambda^2 + \mu_e^2 r^2) \right] R_{-\frac{1}{2}}(r) = 0 \,, \tag{4.112}$$

where λ^2 is a characteristic eigenvalue determined by the condition that $S_{+\frac{1}{2}}(\theta)$ and $S_{-\frac{1}{2}}(\theta)$ are regular at $\theta = 0$ and $\theta = \pi$. We can do the same for the second pair of equation and eliminate $S_{+\frac{1}{2}}(\theta)$ to get an equation for $S_{+\frac{1}{2}}(\theta)$

$$\left[\mathscr{L}_{\frac{1}{2}} \mathscr{L}_{\frac{1}{2}}^\dagger + \frac{a\mu_e \sin\theta}{\lambda + a\mu_e \cos\theta} \mathscr{L}_{\frac{1}{2}}^\dagger + 2(\lambda^2 + a^2 \mu_e^2 \cos^2\theta) \right] S_{-\frac{1}{2}}(\theta) = 0 \,. \tag{4.113}$$

We have thus reduced the solution of the Dirac equation outside a charged, rotating black hole to the solution of a pair of decoupled equations (4.112) and (4.113), from which f_1 follows and accordingly f_2, g_1, g_2 as well as P^B, Q^B. The general solution can be expressed as a linear superposition of the various solutions belonging to the different characteristic eigenvalues of λ^2.

4.5.2 Dirac Equation in Schwarzschild Geometry

Here we discuss the Dirac equation around a Schwarzschild black hole. We follow the derivation given in [8] (for more details see also [5]).

In Schwarzschild geometry we consider only the two radial equations (4.108) and (4.109), which can be rewritten as

$$\Delta^{1/2} \mathscr{D}_0 R_{-\frac{1}{2}} = (1 + i\mu_e r) \Delta^{1/2} R_{+\frac{1}{2}}(r) \,, \tag{4.114}$$

$$\Delta^{1/2} \mathscr{D}_0^\dagger \Delta^{1/2} R_{+\frac{1}{2}} = (1 - i\mu_e r) R_{-\frac{1}{2}}(r) \,, \tag{4.115}$$

where for the orbital angular momentum we restrict here to the case of $l = 1/2$ (in which case the separation constant $\lambda = 1$). The eigenvalue of the angular equation is $(l + 1/2)^2$ since we consider spin $1/2$ particles.

Furthermore, the black hole does not rotate and is not charged; therefore the Boyer–Lindquist coordinates reduce to $\Delta = r^2 - 2Mr$ and $\rho^2 = r^2$, where M is the mass of the black hole. The quantity defined in Eq. (4.99) becomes $K = r^2\sigma$, where σ is the frequency of the incoming Dirac wave. Also in this section we set $G = c = \hbar = 1$. Let us define

$$r_* = r + 2M \log |r - 2M|, \tag{4.116}$$

where $r > r_+ = 2M$, with r_+ the Schwarzschild radius. We see that $r_* \in (-\infty, +\infty)$. We then have

$$\frac{d}{dr_*} = \frac{\Delta}{r^2}\frac{d}{dr}. \tag{4.117}$$

The operators $\mathscr{D}_0$ and $\mathscr{D}_0^\dagger$ take the following forms in terms of r_*

$$\mathscr{D}_0 = \frac{r^2}{\Delta}\left(\frac{d}{dr_*} + i\sigma\right), \tag{4.118}$$

$$\mathscr{D}_0^\dagger = \frac{r^2}{\Delta}\left(\frac{d}{dr_*} - i\sigma\right). \tag{4.119}$$

Furthermore we define $\theta := \arctan(\mu_e r)$, which gives

$$\cos\theta = \frac{1}{\sqrt{1 + \mu_e^2 r^2}}, \tag{4.120}$$

$$\sin\theta = \frac{\mu_e r}{\sqrt{1 + \mu_e^2 r^2}}, \tag{4.121}$$

$$1 \pm i\mu_e r = e^{\pm i\theta}\sqrt{1 + \mu_e^2 r^2}. \tag{4.122}$$

We choose $R_{-\frac{1}{2}} = P_{-\frac{1}{2}}$ and $\Delta^{\frac{1}{2}}R_{-\frac{1}{2}} = P_{\frac{1}{2}}$, which following Chandrasekhar's approach, we write as

$$P_{+\frac{1}{2}} = \psi_{+\frac{1}{2}}\exp\left(-\frac{1}{2}i\arctan(\mu_e r)\right), \tag{4.123}$$

$$P_{-\frac{1}{2}} = \psi_{-\frac{1}{2}}\exp\left(+\frac{1}{2}i\arctan(\mu_e r)\right). \tag{4.124}$$

Define $\hat{r}_* := r_* + \frac{1}{2\sigma}\arctan(\mu_e r)$, which yields

$$d\hat{r}_* = \left(1 + \frac{\Delta}{r^2}\frac{\mu_e}{2\sigma}\frac{1}{1 + \mu_e^2 r^2}\right)dr_*. \tag{4.125}$$

Defining $z_\pm := \psi_{+\frac{1}{2}} \pm \psi_{-\frac{1}{2}}$ the Eqs. (4.114) and (4.115) become

$$\left(\frac{\mathrm{d}}{\mathrm{d}\hat{r}_*} - W\right)z_+ = i\sigma z_- , \tag{4.126}$$

$$\left(\frac{\mathrm{d}}{\mathrm{d}\hat{r}_*} + W\right)z_- = i\sigma z_+ , \tag{4.127}$$

where

$$W = \frac{\Delta^{\frac{1}{2}}(1 + \mu_e^2 r^2)^{\frac{3}{2}}}{r^2(1 + \mu_e^2 r^2) + \mu_e \frac{\Delta}{2\sigma}} . \tag{4.128}$$

Notice that the transformation of the spatial coordinate r to $\hat{r}_*$ shifts the horizon from $r = r_+$ to $\hat{r}_* = -\infty$. From Eqs. (4.126) and (4.127) we get a pair of independent one-dimensional wave equations

$$\left(\frac{\mathrm{d}^2}{\mathrm{d}\hat{r}_*^2} + \sigma^2\right)z_\pm = V_\pm z_\pm , \tag{4.129}$$

where

$$\begin{aligned}
V_\pm &= W^2 \pm \frac{\mathrm{d}W}{\mathrm{d}\hat{r}_*} \\
&= \frac{\Delta^{\frac{1}{2}}(1 + \mu_e^2 r^2)^{\frac{3}{2}}}{[r^2(1 + \mu_e^2 r^2) + \mu_e \frac{\Delta}{2\sigma}]^2}[\Delta^{\frac{1}{2}}(1 + \mu_e^2 r^2)^{\frac{3}{2}} \pm ((r - M)(1 + \mu_e^2 r^2) + 3\mu_e^2 r\Delta)] \\
&\mp \frac{\Delta^{\frac{3}{2}}(1 + \mu_e^2 r^2)^{\frac{5}{2}}}{[r^2(1 + \mu_e^2 r^2) + \mu_e \frac{\Delta}{2\sigma}]^3}\left[2r\left(1 + \mu_e^2 r^2 + 2\mu_e^2 r^3 + \mu_e \frac{r - M}{\sigma}\right)\right] .
\end{aligned} \tag{4.130}$$

Approximate solutions to this equation can be obtained using the WKB method and by imposing boundary conditions at the horizon such that the reflection coefficient vanishes, whereas the transmission coefficient is unity. The parameter space is spanned by the frequency σ and the rest mass of the incoming particle μ_e. It is unphysical[9] when $\sigma < \mu_e$. The rest of the parameter space, i.e., when $\sigma > \mu_e$, is divided into two regions: I-region $E > V_m$ and II-region $E < V_m$, where $V_m = \max(V_+, V_-)$ and $E = \sigma^2$.

In I-region the wave is locally sinusoidal since the wave number k is real for the entire region of $\hat{r}_*$. In II-region the wave is decaying when $E < V$, i.e., where the wave "hits" the potential barrier, while in the rest of the region the wave is propagating.

In I-region Eq. (4.129) is

$$\frac{\mathrm{d}^2 z_+}{\mathrm{d}\hat{r}_*^2} + (\sigma^2 - V_+)z_+ = 0 , \tag{4.131}$$

---

[9] The energy $E \sim \sigma^2$ should be bigger than the rest energy $E = (\mu_e^2 + p^2)^{\frac{1}{2}} \geq \mu_e$.

where $\sigma^2 - V_+ > 0$. This is like a Schrödinger equation, with σ^2 being the total energy of the wave. The solutions are locally sinusoidal, and approximation can be found using the WKB method. Let

$$k(\hat{r}_*) = \sqrt{\sigma^2 - V_+} \, , \tag{4.132}$$

$$u(\hat{r}_*) = \int k(\hat{r}_*) \mathrm{d}\hat{r}_* + \text{const.} \, , \tag{4.133}$$

where k is the wavenumber of the incoming wave and u the *Eiconal*. Then the solution of Eq. (4.131) is approximated as

$$z_+ \sim \frac{A_+}{\sqrt{k}} \exp(iu) + \frac{A_-}{\sqrt{k}} \exp(-iu) \, , \tag{4.134}$$

where $A_+^2 + A_-^2 = k$.

In II-region σ^2 is no longer greater than $V_\pm$ at all radii, and thus the WKB method is not applicable for the whole range of $\hat{r}_*$. As a result $k^2 = \sigma^2 - V_\pm$ may be negative in some regions; therefore k and u are imaginary and the solution (in the "negative" region) is proportional to

$$z_+ \sim \frac{\exp(-\tilde{u})}{\sqrt{\tilde{k}}} \text{ or } \sim \frac{\exp(+\tilde{u})}{\sqrt{\tilde{k}}} \, . \tag{4.135}$$

$\tilde{u}$ and $\tilde{k}$ are the imaginary parts of u and k, respectively. The reflection R and transmission T coefficients can be computed numerically (see [8] for a detailed discussion of the solutions). With these results (radial function, transmission and reflection coefficients) we can study the scattering of $1/2$ spin particles on a Schwarzschild black hole. Sometimes transmission and reflection coefficients are defined at a single point. This is meaningful only if the potential sharply changes in a small region (around the considered point). Furthermore, it has been argued that no bound solutions exist for the Schwarzschild case regardless the value of the mass [9]. This might be due to the fact that there is a finite probability to find the particle inside the horizon.

4.6 The Four Laws of Black Hole Dynamics

There are close analogies between the laws of thermodynamics and the physics of black holes. The reasons behind this are still not fully understood. The first steps in this direction were made by J. Bekenstein and S. Hawking. In 1972, Bekenstein suggested that black holes should have a well-defined entropy and that it is proportional to its event horizon [10, 11]. Two years later, Hawking proposed the existence of blackbody radiation released by black holes due to quantum effects near the event horizon [12]. His work was inspired by a visit to the Soviet scientists Y. Zeldovich and A.

Starobinsky who told him that, according to the Heisenberg uncertainty principle, rotating black holes should create and emit a thermal flux of particles. Hawking radiation is also known as *black hole evaporation* because it reduces the mass and energy of black holes. Because of this, black holes that do not gain mass through other means are expected to shrink and ultimately vanish. In the following we give a short discussion of the four laws of black holes dynamics in analogy to the laws of thermodynamics.

4.6.1   The Zeroth Law

The zeroth law of thermodynamics states that the temperature T throughout a body in thermal equilibrium is uniform. For black holes the analogue of the temperature is the surface gravity κ. The zeroth law of black hole dynamics states that *the surface gravity of a stationary black hole is uniform over the entire event horizon*.

It can be shown[10] that for a Kerr–Newman black hole, the value of the surface gravity κ at the horizon radius r_H is

$$\kappa = \frac{r_H - M}{2Mr_H - Q^2} \, . \tag{4.136}$$

From this relation we see that the surface gravity is constant; therefore Kerr–Newman black holes satisfy the zeroth law of black hole dynamics.

4.6.2   The First Law

The first law of thermodynamics states that the change in the internal energy $U \equiv U(S, V, N, ...)$ is equal to the amount of heat Q supplied to the system, minus the amount of work $\mathcal{W}$ done by the system on its surroundings

$$dU = \delta Q - \delta \mathcal{W} = T\mathrm{d}s - p\mathrm{d}V + \mu\mathrm{d}N + \cdots . \tag{4.137}$$

In black hole physics the analogue of the thermodynamical potential U is the total mass M which is a function of the surface area A of the horizon, the angular momentum J and the electric charge Q.

In order to derive the surface area of the Kerr–Newman horizon we consider its induced metric in Boyer–Lindquist coordinates for fixed t

$$\mathrm{d}s^2 = \rho^2\mathrm{d}\theta^2 + \frac{(r^2 + a^2)^2}{\rho^2} \sin^2\theta\mathrm{d}\phi^2 \, , \tag{4.138}$$

---

[10] See [2] for a detailed derivation of several of the formulas discussed in this chapter.

where

$$r = r_H = M + \sqrt{M^2 - a^2 - Q^2}\,. \tag{4.139}$$

The corresponding volume form is $\bar{\eta} = (r_H^2 + a^2)d\theta \wedge \sin\theta d\phi$, and the surface area of the horizon is

$$A = 4\pi(r_H^2 + a^2) = 4\pi(2Mr_H - Q^2)\,. \tag{4.140}$$

This equation can be solved for the total mass M of the black hole. Using the definition $M_0^2 = \frac{A}{16\pi}$, we find

$$M(A, J, Q) = \left(\left(M_0 + \frac{Q^2}{4M_0}\right)^2 + \frac{J^2}{4M_0^2}\right)^{\frac{1}{2}}\,. \tag{4.141}$$

From this equation we can recover

$$\frac{\partial M}{\partial A} = \frac{1}{8\pi}\kappa\,, \tag{4.142}$$

$$\frac{\partial M}{\partial J} = \Omega_H = \frac{a(2Mr_H - Q^2)}{(r_H^2 + a^2)^2} = \frac{a}{r_H^2 + a^2}\,. \tag{4.143}$$

Before writing down the expression for $\frac{\partial M}{\partial Q}$ we give the electric potential of the black hole

$$\phi = \frac{Q}{\rho^2}r(1 - \Omega_H a \sin^2\theta)\,, \tag{4.144}$$

(see, e.g., [2] for an explicit derivation). At the horizon we have

$$\phi_H = \phi|_H = \frac{Qr_H}{r_H^2 + a^2} = \frac{Qr_H}{2Mr_H - Q^2}\,, \tag{4.145}$$

which is constant, thus

$$\frac{\partial M}{\partial Q} = \phi_H\,. \tag{4.146}$$

Putting all these results together we get a relation for the differential of M, and this is known as the first law of black holes dynamics

$$dM = \frac{1}{8\pi}\kappa\,dA + \Omega_H\,dJ + \phi_H\,dQ. \tag{4.147}$$

The analogy with Eq. (4.137) is striking, which suggests that the temperature T corresponds to the surface gravity κ and the entropy S corresponds to the surface area A of the horizon of the black hole.

Hawking radiation suggests that one should associate a temperature and an entropy to a black hole as follows

$$T_H = \frac{\hbar \kappa}{2\pi k_B} \,, \quad S = \frac{k_B A}{4\hbar} \,, \tag{4.148}$$

where k_B is the Boltzmann constant and $\hbar$ is Planck's constant. The quantity S is known as *Bekenstein–Hawking* entropy and T_H as Hawking temperature. For a complete derivation of these quantities see, e.g., [2].

4.6.3 The Second Law

The second law of black holes dynamics is due to S. Hawking and states that *in any (classical) interaction of matter and radiation with black holes, the total surface area of the boundaries of these holes (as formed by their horizons) can never decrease.* (In the thermodynamics analogy it corresponds to the second law and the concept of entropy.)

The limitation to classical interactions means that we do not consider changes in the quantum theory of matter due to the presence of the strong external gravitational fields of black holes. This is justified for macroscopic black holes, while for "miniholes" quantum effects, such as spontaneous Hawking radiation, would become important. The physics of the latter case requires the use of Dirac equations in curved spacetime.

To provide an example, consider two black holes colliding and coalescing into a single one. The second law tells us that the resulting surface area is larger than the sum of the surface area of the event horizons of both original black holes.

We further investigate this situation assuming that these two black holes are Kerr black holes. Furthermore, let us assume that the final black hole, resulting from the coalescence of the two, is stationary. Due to the second law we have

$$M \left(M + \sqrt{M^2 - a^2} \right) \geq M_1 \left(M_1 + \sqrt{M_1^2 - a_1^2} \right) + M_2 \left(M_2 + \sqrt{M_2^2 - a_2^2} \right) , \tag{4.149}$$

where M_1, M_2 are the masses of the two colliding black holes and M is the mass of the final black hole. We define the *efficiency* as follows:

$$\varepsilon := \frac{M_1 + M_2 - M}{M_1 + M_2} \,. \tag{4.150}$$

As a special case let us take $M_1 = M_2 = \frac{1}{2}\bar{M}$ and $a_1 = a_2 = a = 0$, then $M > \frac{1}{\sqrt{2}}\bar{M}$ (from Eq. (4.149)) and hence

$$\varepsilon \leq 1 - \frac{1}{\sqrt{2}} = 0.293 \,. \qquad (4.151)$$

In principle, a lot of energy can be released as gravitational radiation. It should be stressed that within classical theory the correspondence between temperature and surface gravity is just formal since the physical temperature of a black hole is zero. This is not the case in quantum theory where the physical temperature is given the Hawking temperature $T_H = \frac{\hbar \kappa}{2\pi k_B}$. A test of the area theorem (or second law) has been performed using the first detected gravitational wave event GW150914 [13].

4.6.4 The Third Law

Nernst's third law of thermodynamics states that it is impossible for any process to reduce the entropy of a system to its absolute zero value in a finite number of operations and thus is impossible to bring a system to the absolute zero temperature by any physical process in a finite number of steps. The equivalent law for black holes dynamics states that *the surface gravity of a black hole cannot be reduced to zero within a finite advanced time*.

4.7 Hawking Radiation

Energy emission is possible from a static black hole, provided quantum effects are taken into account. Indeed, Hawking radiation indicates the phenomenon of particle creation by black holes, which arises when considering quantum field theory in a Schwarzschild or Kerr spacetime. In quantum field theory on a curved spacetime, the metric is treated classically but is coupled to matter fields which are treated quantum mechanically, like in the well-known case of quantum field theory in an external classical potential. If one considers a spacetime which has an initial flat region, followed by a region of curvature and by a final flat region, this theory can be treated like a scattering theory. Let us assume that a pair of particle and antiparticle is created through vacuum quantum fluctuations, where one is created inside and the other one outside the event horizon of the black hole. In this scenario the latter one can reach a distant observer with energy $E > 0$.

A detailed treatment of the phenomenon requires quantum field theory in curved spacetime. Here we give just a brief description, which is of course not exhaustive (for a more detailed treatment see, e.g., [14, 15]). In order to make the analysis easier we consider a scalar field and thus use the Klein–Gordon equation rather than the Dirac equation. The Klein–Gordon equation reads

$$(\Box_g + \mu_p^2)\phi = 0 \,, \qquad (4.152)$$

where μ_p is the mass of the scalar field of the particle p and $\Box_g$ is the d'Alembert operator for the metric g

$$\Box_g = \frac{1}{|g|} \partial_\nu \left(\sqrt{|g|} g^{\mu\nu} \partial_\mu \right) . \tag{4.153}$$

One has to consider the formalism of second quantization and interpret ϕ as an operator

$$\phi = \sum_i (F_i a_i + \bar{F}_i a_i^\dagger), \tag{4.154}$$

where F_i are a complete orthonormal family of complex valued positive frequency solutions of the Klein–Gordon equation, while $\bar{F}_i$ are their complex conjugates (which represent negative frequency solutions). Indeed, the scattering of these solutions by the classical metric (of a curved spacetime) will cause a mixing between the sets F_i and $\bar{F}_i$, which results in a different splitting of the operator ϕ into annihilation and creation operators. This means that the initial vacuum state in the initial flat region will not be the same as the final vacuum state in the final flat region. This also implies that, in order to have a well-defined notion of a particle, one has to be able to define the notion of positive and negative frequency unambiguously, which is possible in a flat spacetime but not, for example, on the black hole horizon. Therefore, it is not clear what the physical meaning of a "particle at the horizon" is.

We consider a Schwarzschild black hole for which we introduce Regge–Wheeler coordinates $(t, r_*, \theta, \varphi)$. Using natural units, i.e., $G = c = \hbar = 1$, r_* is given by

$$r_* = r + 2M \log \left(\frac{r}{2M} - 1 \right) , \tag{4.155}$$

with t, θ, φ fixed, and $r > 2M$ for particles outside the black hole. This function maps $r \in \,]2M, \infty[\, \mapsto r_* \in \,] -\infty, \infty[$. Since

$$\frac{\mathrm{d}r_*}{\mathrm{d}r} = 1 + \left(\frac{r}{2M} - 1 \right)^{-1} = \left(1 - \frac{2M}{r} \right)^{-1} , \tag{4.156}$$

the metric, with $r = r(r_*)$, reads

$$\mathrm{d}s^2 = \left(1 - \frac{2M}{r} \right) (\mathrm{d}t^2 - \mathrm{d}r_*^2) + r^2 (\mathrm{d}\theta^2 + \sin^2\theta \mathrm{d}\varphi^2) . \tag{4.157}$$

We write the Klein–Gordon equation in Regge–Wheeler coordinates, after separating the angular part according to

$$f(t, r, \theta, \varphi) = \sum_{l=0}^{\infty} \sum_{m=-l}^{l} \frac{f_{lm}(t, r)}{r} Y_{lm}(\theta, \varphi) , \tag{4.158}$$

We get

$$\left(\frac{\partial^2}{\partial^2 t} - \frac{\partial^2}{\partial^2 r_*} + V_l \right) f_{lm} = 0 \,, \tag{4.159}$$

where V_l is the effective potential given as

$$V_l(r_*) = \left(1 - \frac{2M}{r} \right) \left(\frac{2M}{r^3} + \frac{l(l+1)}{r^2} + \mu_p^2 \right) \,, \tag{4.160}$$

and has limits

$$V_l(r_*) \rightarrow \begin{cases} 0, & r_* \rightarrow -\infty \,, \text{i.e. } r \rightarrow 2M \\ \mu_p^2, & r_* \rightarrow \infty \,, \text{i.e. } r \rightarrow \infty \end{cases} \,. \tag{4.161}$$

Equation (4.159) has precisely the form of the wave equation for a massless scalar field in a two-dimensional flat spacetime (with Cartesian coordinates t, r_*) with scalar potential (4.160).

For $r_* \rightarrow -\infty$, i.e., $r \rightarrow 2M$, the solutions, following the scattering approach, are of the form

$$f_{lm}(t, r_*) = f_{out}(t - r_*) + f_{in}(t + r_*) \,, \tag{4.162}$$

where f_{in} describes the incoming wave from the white hole[11] and f_{out} the outgoing wave from the black hole.

4.7.1 Expected Number of Outgoing Particles

We do the analysis of the particle creation effect using the theory of scattering by potentials in flat spacetime, as it is done, e.g., in particle physics.

We assume $\mu_p \neq 0$, then every wave packet should behave, in the asymptotic past, like a free (i.e., $V = 0$) massless solution in (t, r_*)-space, propagating in from $r_* \rightarrow -\infty$ (corresponding to the white hole horizon). To this solution we have to include a massive solution (distorted by a $\frac{1}{r_*}$ potential[12]) propagating in from $r_* \rightarrow \infty$. Similarly, in the asymptotic future, every wave packet should behave as a free massless wave propagating to $r_* \rightarrow -\infty$ together with a (distorted) massive wave propagating to $r_* \rightarrow \infty$. For $\mu_p = 0$, we expect that every wave packet approaches a free massless solution in both the asymptotic past and future. Near the horizon, for $r_* \rightarrow -\infty$ (i.e., $r \rightarrow 2M$), the "free wave" ($\sim e^{-i\omega t}$) looks like

---

[11] In general relativity a white hole is a hypothetical region of spacetime which cannot be entered from the outside, although matter and light can escape from it. In this sense, it is the reverse of a black hole, which can only be entered from the outside and from which matter and light cannot escape.

[12] Equation (4.160) behaves as $\sim (\mu_p^2 - 2M \mu_p^2 / r_*)$ for $r_* \rightarrow \infty$.

$$\sim a\exp(-i\omega u) + b\exp(-i\omega v)\,, \tag{4.163}$$

as given by Eq. (4.159), where $\omega > \mu_p$, $u = t - r_*$ and $v = t + r_*$. The case of a purely outgoing solution at the horizon is given by $b = 0$. It is found (see, e.g., [14]) that each outgoing mode of frequency ω has an "infinite oscillation" singularity on the black hole horizon. The behavior of these solutions is

$$\sim a\exp\left(\frac{i\omega}{k}\ln(-U)\right)\,, \tag{4.164}$$

where $U = -e^{-ku}$ is a Kruskal coordinate with $k = \frac{1}{4M}$. Next, one finds that the time dependence of ϕ, the solution of the Klein–Gordon equation, for $v = 0$ has the form

$$\phi(v) = \begin{cases} 0\,, & v > 0 \\ \phi_0 \exp\left(\frac{i\omega}{k}\ln(-\alpha v)\right)\,, & v < 0 \end{cases}\,, \tag{4.165}$$

where $\alpha \sim \frac{dU}{d\lambda}\big|_{\lambda=0}$ (λ is the affine parameter describing a geodesic entering the black hole, with $\lambda = 0$ chosen to correspond to the intersection between the geodesic and the horizon and $U(\lambda = 0) = 0$ on the event horizon. Furthermore since λ depends smoothly on U and satisfies $\frac{dU}{d\lambda} \neq 0$ on the horizon, it follows that near $\lambda = 0$, each outgoing mode oscillates as a function of λ as $\alpha \sim \exp(i\omega k^{-1}\ln(-\alpha\lambda))$.

It has to be noted that even when starting from purely positive frequencies for ω, ϕ also contains negative frequencies (which can be interpreted as antiparticles). When performing the Fourier transform of ϕ with respect to v one finds that $\hat{\phi}(-\sigma) = -exp(-\pi\omega/k)\hat{\phi}(\sigma)$. Thus the magnitude of the negative frequency part is $exp(-\pi\omega/k)$ times the magnitude of the positive frequency part.

Next, one can show that the number of particles spontaneously created by such a wave packet is given by ($c = 1$)

$$\langle N\rangle = |t|^2 \frac{e^{-2\pi\omega/k}}{1 - e^{-2\pi\omega/k}}\,, \tag{4.166}$$

where $|t|^2$ is the square of Klein–Gordon norm of the wave packet reaching infinity. Equation (4.166) corresponds to the formula which holds for a perfect black body emitter at a temperature given by[13]

$$E = k_B T = \frac{\hbar k}{2\pi c} = \frac{\hbar c^3}{8\pi G M}\,, \tag{4.167}$$

where $k = \frac{c^4}{4\pi M G}$ and

$$T \simeq 6\cdot 10^{-8}\frac{M_\odot}{M}\ \mathrm{K}\,. \tag{4.168}$$

---

[13] Note that $\frac{h\nu}{k_B T} \overset{!}{=} \frac{2\pi\omega}{ck}$.

Equation (4.167) gives us the *Hawking temperature* as

$$T_H = \frac{\hbar c^3}{8\pi \, GMk_B} \, .$$ (4.169)

4.7.2 Black Hole Evaporation

We perform a crude analytic estimate of the radiated power. We assume that only photons are radiated away and use the Stefan–Boltzmann law of blackbody radiation. The Stefan–Boltzmann constant is given by $\sigma = \frac{\pi^2 k_B^4}{60\hbar^3 c^2}$, and the Schwarzschild radius is $r_s = \frac{2GM}{c^2}$. The Schwarzschild sphere surface of the black hole is

$$A_s = 4\pi r_s^2 = 4\pi \left(\frac{2GM}{c^2}\right)^2 = \frac{16\pi \, G^2 M^2}{c^4} \, .$$ (4.170)

To obtain our rough estimate of the emitted power we use the Stefan–Boltzmann power law $P = A_s \sigma \varepsilon T^4$ assuming the black hole to be a perfect blackbody $\varepsilon = 1$

$$P = A_s \sigma T_H^4 = \left(\frac{16\pi \, G^2 M^2}{c^4}\right)\left(\frac{\pi^2 k_b^4}{60\hbar^3 c^2}\right)\left(\frac{\hbar c^3}{8\pi \, GMk_B}\right)^4 = \frac{\hbar c^6}{15360\pi \, G^2 M^2} = \frac{K_{ev}}{M^2} \, ,$$ (4.171)

where $K_{ev} \simeq 3.5 \times 10^{32}$ (W kg^2). Under the assumption of an otherwise empty universe, it is possible to calculate how long it would take for a black hole to evaporate. We relate the rate of evaporation energy loss of the black hole

$$P = -\frac{dE}{dt} = \frac{K_{ev}}{M^2} \, ,$$ (4.172)

to its mass loss using Einstein's mass–energy relation $E = Mc^2$

$$P = -\frac{dE}{dt} = -c^2 \frac{dM}{dt} \, ,$$ (4.173)

and get through integration

$$\int_{M_0}^{0} M^2 dM = -\frac{K_{ev}}{c^2} \int_0^{t_{ev}} dt \, ,$$ (4.174)

where M_0 is the initial black hole mass and t_{ev} the *evaporation time*. Solving this equation with respect to t_{ev} gives

$$t_{ev} = \frac{c^2 M_0^3}{3 K_{ev}} = \left(\frac{c^2 M_0^3}{3} \right) \left(\frac{15360\pi G^2}{\hbar c^6} \right) = \frac{5120\pi G^2 M_0^3}{\hbar c^4} = 8.4 \times 10^{-17} \left(\frac{M_0}{\text{kg}} \right)^3 \text{s}.$$

(4.175)

For $M_0 = M_\odot$ we get

$$t_{ev} = \frac{5120\pi G^2 M_\odot^3}{\hbar c^4} = 6.6 \times 10^{74} \simeq 2.1 \cdot 10^{67} \text{ yr},$$

(4.176)

This is a huge number as compared to the age of the universe $t_{universe} \simeq 1.4 \times 10^{10}$ year.

4.8   Summary

It took almost 50 years after the publication by Einstein of his theory on general relativity to find the vacuum solution for a rotating black hole. We have briefly discussed the Kerr and the Kerr–Newman solutions of Einstein's vacuum field equations. As next we introduced the tetrad formalism and discussed some concepts of differential geometry. We then treated spinors in curved spacetime, for which the tetrad formalism is useful. In the following we derived the Dirac equation in curved spacetime and discussed its solutions around a Schwarzschild black hole. As a last topic we presented the four laws on black hole dynamics and Hawking radiation.

4.9   Problems

Problem 4.1   (*Object falling into a black hole*)
   An object with a length of about 2 m and mass μ is dropped radially into a solar mass black hole from constant spatial coordinates (r_*, θ_*, ϕ_*).

(a) Treating the in-falling object as a point particle, when calculating its coordinate time as a function of r, it shows that the coordinate speed

$$\frac{dr}{dt} = \frac{\dot{r}}{\dot{t}} = -B(r) \sqrt{1 - \frac{B(r)}{B(r_*)}},$$

becomes zero as the object approaches the horizon. Thus from the outside one will never see the object falling into the black hole.

(b) Let us take into account that the in-falling object is not a point particle, but rather an object with length of about 2 m, which we assume to break apart if the tidal acceleration overcomes 20 g, g being the Earth's gravitational acceleration at the surface. Calculate the distance at which the object breaks apart using the equation of geodesic deviation. Choose an orthonormal frame (tetrad) parallel transported along the geodesic and use Cartan's structure equations to calculate the connection and curvature forms.

(c) If the object falls instead into a supermassive black hole like the one in the center of our galaxy ($M \approx 4 \times 10^6 M_\odot$), where will it break apart? Before or after passing the event horizon?

References

1. R. Kerr, Phys. Rev. Lett. **11**, 237 (1963)
2. N. Straumann, *General Relativity* (Springer, 2013)
3. C. Misner, K. Thorne, J. Wheeler, *Gravitation* (Freeman, 1973)
4. S. Weinberg, *Gravitation and Cosmology* (Wiley, 1972)
5. S. Chandrasekhar, *The Mathematical Theory of Black Holes* (Oxford, 1983)
6. S. Chandrasekhar, The solution of Dirac's equation in Kerr geometry. Proc R Soc London Ser A **349**, 571–575 (1976)
7. D.N. Page, Dirac equation around a charged, rotating black hole. Phys. Rev. D **14**, 1509–1510 (1976)
8. B. Mukhopdhyay, S.K. Chakrabarti, Semi-analytical solution of Dirac equation in Schwarzschild geometry. Class. Quantum Gravity **16**, 3165–3181 (1999)
9. D. Batic, M. Nowakowski, K. Morgan, The problem of embedded eigenvalues for the Dirac equation in the Schwarzschild black hole metric. Universe **2**, 31 (2016)
10. J.D. Bekenstein, Black holes and the second law. Lettere al Nuovo Cimento **4**, 737–740 (1972)
11. J.D. Bekenstein, Black holes and entropy. Phys. Rev. D **7**, 2333 (1973)
12. S.W. Hawking, Black hole explosions? Nature **248**, 30–31 (1974)
13. M. Isi et al., Testing the Black-Hole Area law with GW150914. Phys. Rev. Lett. **127**, 011103 (2021)
14. R. M. Wald, *General Relativity* (University of Chicago Press, 1984)
15. R.M. Wald, *Quantum field theory in curved spacetime and black holes thermodynamics* (University of Chicago Press, 1994)

Chapter 5
Tests of General Relativity

In this last chapter we discuss the Einstein equivalence principle, which can be reformulated in three principles: the weak equivalence principle, the local Lorentz invariance and the local position invariance. We briefly present also Schiff's conjecture and the strong equivalence principle. The different assumptions of the principle of equivalence (EP), which were the starting points for the formulation of the general theory of relativity, first written down by Albert Einstein in his famous work "Grundlagen der allgemeinen Relativitätstheorie" [1] in 1916, have been confirmed by a broad range of experiments to a very accurate level, and their further confirmation is still an active field of research.

5.1 The Weak Equivalence Principle

There exist a variety of different formulations and types of EP which contain different assumptions and are therefore referred to as "weaker" or "stronger" (see, e.g., [2, 3]). Here, we mainly focus on the weak, the Einstein and the strong equivalence principle.

The weak equivalence principle (WEP) or universality of free fall (UFF) states that all test bodies with negligible self-gravity behave the same in a gravitational field, meaning they experience the same acceleration, independent of their internal structure or composition. H. Bondi named the terms (matter and weight) inertial mass m_I and (passive) gravitational mass m_G[1] [4], so that the principle of equivalence can be written as

$$m_I = m_G. \tag{5.1}$$

---

[1] The passive gravitational mass denotes the mass that gets affected by gravity, in contrast to the active gravitational mass, which creates the gravitational field.

© The Author(s), under exclusive license to Springer Nature Switzerland AG 2022
P. Jetzer, *Applications of General Relativity*, UNITEXT for Physics,
https://doi.org/10.1007/978-3-030-95718-6_5

A more precise definition of the WEP as given, e.g., by C. Will [2] is the following: "If an uncharged test body is placed at an initial event in spacetime and given an initial velocity there, then its subsequent trajectory will be independent of its internal structure and composition." An uncharged test body in this context means an electrically neutral body with negligible self-gravitational energy. There is a parameter σ to define when self-gravity is negligible [3]

$$\sigma = \frac{Gm}{c^2 r}, \tag{5.2}$$

where G is the gravitational constant, m is the mass of the test body (it does not matter if gravitational or inertial mass, since they are assumed to be equal), c is the speed of light and r is the size of the test body. For example an atom can be regarded as a test body, since it is uncharged and the self-gravitational energy is negligible as $\sigma \sim 10^{-43}$.

5.2   The Einstein Equivalence Principle

Einstein stated that if all the bodies fall with the same acceleration in an external gravitational field, then in a freely falling system, the mechanical laws will behave as if gravity were absent. He even added that not only mechanical laws, but all the laws of physics should behave this way. Einstein used the principle of equivalence as a basic element in the development of general relativity. Around 1960, R. Dicke contributed with crucial ideas about the foundations of gravitation theory [5, 6] and generalized the principle to what today is called the *Einstein equivalence principle* (EEP).[2] The EEP consists of the following three assumptions, which may appear as completely independent statements in the first place.

1. The *weak equivalence principle* (WEP), which was introduced above, is valid, and therefore all test bodies experience the same acceleration in a gravitational field. This assumption is essential for the existence of any local freely falling frames.
2. *Local Lorentz Invariance* (LLI): The outcome of any local[3] non-gravitational experiment is independent of the velocity of the freely falling reference frame in which it is performed.
3. *Local Position Invariance* (LPI): The outcome of any local non-gravitational experiment is independent of where and when in the universe it is performed.

The last two points therefore state that there do neither exist any preferred frame nor location effects. It is possible to argue convincingly that if EEP is valid, then the effects of gravity must be equivalent to the effects of living in a curved spacetime.

---

[2] For a comprehensive review we refer to [2, 7].

[3] A local test experiment in this context means that it is small, such that there exist no inhomogeneities in the gravitational potential; see[8].

As a consequence, the only theories of gravity that can fully embody EEP are those that satisfy the postulates of "metric theories of gravity" which are:

1. Spacetime is endowed with a symmetric metric.
2. The trajectories of freely falling test bodies are geodesics of that metric.
3. In local freely falling reference frames, the non-gravitational laws of physics are those written in the language of special relativity.

Indeed, if EEP holds, then in local freely falling frames the laws governing experiments must be independent of the velocity of the frame (LLI), with constant values for the various atomic constants (in order to satisfy LPI). The only known laws that fulfill this are the ones which are compatible with special relativity, such as Maxwell's equations of electromagnetism.

It can be shown (see, e.g., [2]) that if all the three pillars and therefore the EEP are valid, the theory of gravity must satisfy the postulates of metric theories of gravity and can therefore be called a metric theory. This contains the following conditions: first that spacetime is endowed with a symmetric metric, which in the case of GR would be the metric tensor $g_{\mu\nu}$, and second that the trajectories of freely falling test bodies are geodesics of that metric. A test body in a gravitational field described by the metric $g_{\mu\nu}(x)$, which may depend on coordinates x and therefore correspond to a curved spacetime, fulfills the following equation of motion (see Eq. 1.43)

$$\frac{\mathrm{d}^2 x^\kappa}{\mathrm{d}\tau^2} = -\Gamma^\kappa_{\mu\nu} \frac{\mathrm{d}x^\mu}{\mathrm{d}\tau} \frac{\mathrm{d}x^\nu}{\mathrm{d}\tau}, \tag{5.3}$$

where τ and $\Gamma^\kappa_{\mu\nu}$ are the proper time and the Christoffel symbols, respectively, which are connected to the metric by the following expression (see Eq. 1.7)

$$\Gamma^\kappa_{\mu\nu} = \frac{g^{\kappa\lambda}}{2} \left(\frac{\partial g_{\lambda\nu}}{\partial x^\mu} + \frac{\partial g_{\lambda\mu}}{\partial x^\nu} - \frac{\partial g_{\mu\nu}}{\partial x^\lambda} \right). \tag{5.4}$$

The solution of the equation of motion of the test body leads to geodesics of $g_{\mu\nu}(x)$. The third and last postulate is that in local freely falling reference frames, the non-gravitational laws of physics are those written in the language of special relativity. The EEP can therefore be regarded as the foundation of all curved spacetime or metric theories of gravity, which means that all non-gravitational fields couple in the same way to the gravitational field. If this is satisfied, the theory of gravitation must be a phenomenon of curved spacetime.

The three parts of the EEP, namely the WEP, LLI and LPI, are usually tested independently of each other. However, we will see that if Schiff's conjecture is correct, certain experiments yield confirmations on more than one part of the EEP at the same time.

5.2.1 Tests of the Weak Equivalence Principle

The WEP can be tested by so-called Eötvös experiments, in which the acceleration due to gravity of two different test bodies is compared. If WEP is valid, every test body, independent of their internal structure or composition, should experience the same acceleration in vacuum. A possible violation of WEP can be introduced, by allowing a certain form of energy A (e.g., electromagnetic, hyperfine energy, etc.) to contribute differently to the gravitational mass m_G than it does to the inertial mass m_I by

$$m_G = m_I + \sum_A \frac{\eta^A E^A}{c^2}, \tag{5.5}$$

where E^A is the amount of energy of form A and η^A is a dimensionless parameter which measures the strength of its violation of the WEP. The acceleration of a test body with gravitational mass m_G and inertial mass m_I under influence of a gravitational field is given by using Newton's law

$$a = \frac{m_G}{m_I} g = \left[1 + \sum_A \eta^A \left(\frac{E^A}{m_I c^2} \right) \right] g, \tag{5.6}$$

where g is the gravitational acceleration. For a test body X with a negligible amount of self-gravitational energy, its acceleration in a gravitational field can be written as

$$a_X = \left(1 + \sum_A \eta^A \zeta_X^A \right) g, \tag{5.7}$$

where the fractional energy contribution of interaction A to the total energy of a test body X is introduced as

$$\zeta_X^A = \frac{E^A}{m_I c^2} \bigg|_X . \tag{5.8}$$

To measure the violation of the WEP in experiments, the Eötvös ratio η, which is a measure of the difference in acceleration of two test bodies X and Y experience in a gravitational field, is used. It is defined as

$$\eta(X, Y) \equiv \frac{2(a_X - a_Y)}{a_X + a_Y}. \tag{5.9}$$

Substituting the expression for the acceleration of the test body X, which allows a violation of the WEP, given in Eq. (5.7) into the definition of the Eötvös ratio yields

$$\eta(X, Y) = \frac{2 \sum_A \eta^A \left(\zeta_X^A - \zeta_Y^A \right)}{2 + \sum_A \eta^A \left(\zeta_X^A + \zeta_Y^A \right)}. \tag{5.10}$$

As we know by experimental confirmation the violation of WEP is very small $\eta^A \ll 1$, the second term in the denominator can thus be neglected and we get the approximation

$$\eta(X, Y) \simeq \sum_A \eta^A \left(\zeta_X^A - \zeta_Y^A \right). \tag{5.11}$$

If we assume that the violations of the WEP by the different forms of energy do not compensate each other or that only one single form of energy A couples non-metrically to gravity and thus violates the WEP, we can deduce the following relation between the Eötvös ratio and the WEP violation parameter η^A

$$|\eta(X, Y)| \geq \left| \eta^A \left(\zeta_X^A - \zeta_Y^A \right) \right|, \tag{5.12}$$

from which an expression for the upper limit on the strength of violation of the WEP by energy of form A

$$|\eta^A| \leq \frac{|\eta(X, Y)|}{\left| \zeta_X^A - \zeta_Y^A \right|}, \tag{5.13}$$

can be deduced. Limits on the Eötvös ratio directly lead to limits on the strength of violation of the WEP. It is important to notice that the value of η depends on the experimental settings used (e.g., the internal structure and composition of the test bodies or the object used as an attractor). These limits are therefore not universal but depend mainly on the test bodies used. The limits placed on the WEP violation parameter η^A however are universal and can be directly compared to each other. If the WEP is valid, $\eta^A = 0$ which leads to $\eta = 0$, finding lower bounds on η and especially on the different η^A's supports the confirmation of the WEP.

A restriction on the violation of the WEP by $|\eta| \leq 10^{-3}$ for various substances was already found in the seventeenth century. Indeed, Galileo Galilei tested the UFF by using masses of different compositions bound to wires of the same length and comparing how long the pendulums kept step with each other. Similarly, Newton performed different pendulum experiments which are reported in his *Principia*.

R. von Eötvös, D. Pekár and E. Fekete could improve the upper bound to $|\eta| \leq 5 \times 10^{-9}$ at the beginning of the twentieth century, by using the Earth as an attractor and rotating the apparatus about the direction of the wire [9].

The best limits obtained by experiments performed on Earth are the ones obtained by the so-called Eöt-Wash experiments performed at the University of Washington, whose most stringent limits are given by $|\eta(\text{Be}, \text{Ti})| \leq 2.1 \times 10^{-13}$ and $|\eta(\text{Be}, \text{Al})| \leq 2 \times 10^{-13}$, respectively [10]. The MICROSCOPE satellite , launched on April 25, 2016, was put in a Sun-synchronous circular orbit around the Earth at an altitude of 710 km, with the aim to test WEP. It yielded already in December 2017 an upper limit to $|\eta(\text{Ti}, \text{Pt})| \leq 1.3 \times 10^{-14}$ [11]. This is the lowest value up to date, but the mission aims to improve the upper bound to one part in 10^{15}, when more orbits and more data analysis will have been done. Future space experiments, either improving the technology used by MICROSCOPE or by using new methods based on atom interferometry will further improve the accuracy [12, 13]. The "Antimatter Experiment: Gravity, Interferometry, Spectroscopy" (AEgIS) experiment tries to confirm the WEP with antimatter [14].

5.2.2 Tests of Local Lorentz Invariance

Special relativity has been tested in depth; see for example particle physics experiments. In the past several years, a vigorous theoretical and experimental effort has been launched to find violations of special relativity to look for evidence of new physics "beyond" Einstein, such as violations of Lorentz invariance that might result from certain models of quantum gravity. Quantum gravity asserts that there is a fundamental length scale given by Planck length $l_p = (\frac{\hbar G}{c^3})^{\frac{1}{2}} = 1.6 \times 10^{-33}$ cm, but since length is not an invariant quantity, there could be a violation of Lorentz invariance at some level in quantum gravity.

A useful way to interpret some experiments is to suppose that the electromagnetic interactions suffer a slight violation of Lorentz invariance, through a change in the speed of electromagnetic radiation c relative to the limiting speed of material test particles, in other words $c \neq 1$. Such a violation necessarily selects a preferred universal rest frame. Such a Lorentz non-invariant electromagnetic interaction would cause shifts in the energy levels of atoms and nuclei that depend on the orientation of the quantization axis of the state relative to our universal velocity vector and on the quantum numbers of the state.

The results of these experiments are usually given in terms of the quantity $\delta :=$ $|\frac{1}{c^2} - 1|$. Astrophysical observations have also been used to put bound on Lorentz violations.

If there is a violation of LLI of any form of energy A, one could expect a contribution to the inertial mass δm_I^{ij}, called anomalous inertial mass tensor. It is of the form

$$\delta m_I^{ij} \sim \sum_A \delta^A \frac{E^A}{c^2}, \tag{5.14}$$

where δ^A is a dimensionless parameter for the strength of anisotropy induced by interaction A. A violation of LLI leads to preferred frame effects. A preferred frame exists which could possibly be the cosmic microwave background (CMB). The bare violation of LLI can be written as

$$\delta_0^A = \left(\frac{c}{w}\right)^2 \delta^A, \tag{5.15}$$

where w is the velocity of the laboratory with respect to the preferred frame. In the case of the CMB $w/c = 1.23 \times 10^{-3}$ or $w = 369$ km/s [15]. Limits on the strength of LLI violation can therefore be inferred from limits on δm_I^{ij} by assuming, as in the case of the WEP, that only one form of energy violates LLI, or at least that the different violations do not compensate each other

$$|\delta^A| \le \frac{|\delta m_I^{ij} c^2|}{|E^A|}. \tag{5.16}$$

By dividing Eq. (5.14) by the mass of the test body X, we get an expression including the fractional energy contribution ζ_X^A instead of the energy E^A

$$\frac{\delta m_I^{ij}}{m_X} = \sum_A \delta^A \zeta_X^A. \tag{5.17}$$

This will be useful to relate violations of the WEP and LLI.

For the case of the electromagnetic interaction a formalism has been developed which allows a deviation from the speed of light $c \ne c_0$, where c_0 is its universal value [16]. This method yields the following expression for the anomalous inertial mass tensor [7]

$$\delta m_I^{ij} c^2 = -2\delta \left[\frac{4}{3} E^{ES} \delta^{ij} + \left(E^{ES} \right)^{ij} \right], \tag{5.18}$$

where $E^{ES} \delta^{ij}$ contributes to the isotropic and $\left(E^{ES} \right)^{ij}$ to the anisotropic part of the anomalous inertial mass tensor. The dimensionless parameter δ is defined as

$$\delta \equiv \left(\frac{c_0}{c} \right)^2 - 1, \tag{5.19}$$

which is equal to zero if LLI is valid.

To include all the interactions of the standard model a framework called the standard model extension (SME) has been developed [17, 18]. The SME inserts a variety of tensorial quantities to the terms in the action of the standard SU(3) × SU(2) × U(1) field theory of particle physics. Experimental bounds on all the various parameters of the SME can be found in [19]. For example, an upper bound to $|\delta m_{I,\text{aniso}}^{ij} c^2| \le 6.7 \times 10^{-25}$ eV by comparing the energy splitting in ^{3}He and ^{129}Xe was found [20]. All these experiments set very stringent limits on the anisotropic parts of δm^{ij}. To test the scalar part of δm^{ij} another experiment is needed.

5.2.3 Tests of Local Position Invariance

The principle of LPI can be tested by a gravitational redshift experiment using clocks. Every type of clock depends either on oscillation frequencies between energy states or on the decay rate of a compound. This means that every type of clock depends on the transition of a certain form of energy and can therefore be used to test for a possible violation of LPI by this form of energy. The most common experiment to test LPI is to set two clocks in different gravitational potentials and comparing their ticking rates, to the one predicted by gravitational redshift of GR. This is referred to as the standard gravitational redshift experiment.

This shift in frequency, referred to as gravitational redshift z, between two identical clocks at rest at different heights in a static gravitational field, is given by

$$z = \frac{\Delta \nu}{\nu} = -\frac{\Delta \lambda}{\lambda}. \tag{5.20}$$

This frequency shift can also be written as

$$z = \frac{\Delta U}{c^2}, \tag{5.21}$$

where ΔU is the difference in the Newtonian gravitational potential between the two positions of the clocks. If we allow a deviation from the gravitational redshift, it can be written as

$$z = (1 + \alpha) \frac{\Delta U}{c^2}, \tag{5.22}$$

where α is a dimensionless parameter which measures the strength of the violation of LPI.

First high precision measurements were done by the Pound–Rebka–Snider experiments between 1960 and 1965 using γ-ray photons from ^{53}Fe as they ascended and descended the Jefferson Physical Laboratory Tower at Harvard [21, 22].

One of the most precise standard redshift tests to date was the Vessot–Levine rocket experiment known as Gravity Probe-A (GPA) that took place in June 1976 [23]. A hydrogen maser was flown on a rocket to an altitude of 10'000 km and its frequency compared to a hydrogen maser on the ground. The analysis of the data yielded an upper limit of $|\alpha| < 2 \times 10^{-4}$. Using the opportunity given by the Galileo 5 and 6 satellites, as they were wrongly put in a slightly eccentric orbit, it was possible to improve the above limit to $|\alpha| < (3 - 4) \times 10^{-5}$ [24]. A similar limit has also been found using a pair of portable clocks put at different heights on Earth [25].

The *Atomic Clock Ensemble in Space* (ACES) by ESA, scheduled to be launched earliest in 2023, is going to bring on the International Space Station (ISS) an advanced hydrogen maser and a cold atom clock based on cesium (PHARAO). An improvement of the measurements of two orders of magnitude is expected, which will bring the upper limit to $\alpha < 10^{-6}$.

If the LPI is satisfied, then the fundamental constants of non-gravitational physics should be constants in time. More recent experiments have used strontium-87 or rubidium-87 atoms trapped in optical lattices and compared with cesium-133 to obtain

$$\frac{\dot{\alpha}_{EM}}{\alpha_{EM}} < 10^{-16} \, \text{year}^{-1}, \tag{5.23}$$

where $\alpha_{EM} = \frac{e^2}{\hbar c} \simeq \frac{1}{137}$ is the fine structure constant.

Future gravitational redshift experiments may be able to test possible deviations from the Newtonian potential up to the order of v^2/c^2. The gravitational potential can then be approximated by

$$U(r) = -1 + U_0(r) + U_1(r) = -1 - \frac{2Gm}{r} + \frac{2}{c^2}\left(\frac{Gm}{r}\right)^2, \qquad (5.24)$$

which, by inserting this approximation in the expression given in equation (5.22), could give rise to new LPI violation parameters α_0 and α_1 that could be tested separately. However, the last term U_1 is many orders of magnitude smaller than U_0, e.g., for the experimental settings of the ACES mission one can approximate $|\Delta U_1/\Delta U_0| \simeq 10^{-9}$ or $|\Delta U_1/c^2| \simeq 10^{-19}$. The accuracy of ACES is most likely not high enough to detect this term, since it is currently at a level of parts in 10^{-16}.

A different type of experiments used to test LPI is the so-called null redshift experiments. These test different types of clocks at the same position and look if their relative rates depend upon the gravitational potential. For example the comparison of two hydrogen maser clocks and three superconducting cavity-stabilized oscillator (SCSO) clocks yielded the limit [26]

$$|\alpha_H - \alpha_{SCSO}| \le 1.7 \times 10^{-2}, \qquad (5.25)$$

whereas the comparison between hyperfine transitions in ^{87}Rb and ^{133}Cs yielded [27]

$$|\alpha_{Rb} - \alpha_{Cs}| \le 1.2 \times 10^{-6}. \qquad (5.26)$$

If LPI is valid, the laws of physics are independent not only of the position, but of time, too. This leads to the constraint that non-gravitational constants must be constant in time.

5.3 Schiff's Conjecture

In 1960, Leonard I. Schiff conjectured that the three tests of the EEP are not independent of each other [28]. The confirmation of the WEP is by the same time a confirmation of LLI and LPI, and one might say that the WEP implies the EEP and the metric postulate. Or stated differently, if WEP is valid, the theory of gravitation must be a metric one. A rigorous proof of the conjecture is impossible (some counterexamples are known [29]), yet a number of powerful "plausibility" arguments can be formulated (see, e.g., [16]). The most general ones are based on the assumption of energy conservation, which was already used in this context by Einstein as an argument for the existence of the gravitational redshift.

Following on a cyclic thought experiment similar to the one mentioned by Nordtvedt [30], we will get a quantitative relationship between violations of the

WEP and LPI, or more precisely, between Eötvös and gravitational redshift experiments. Similarly, again based on a thought experiment, one can also find relations between violations of WEP and LLI (see Haugan [31]). For more details we refer to the review of Will [7].

5.3.1 Quantitative Relationship Between Violations of the Weak Equivalence Principle and Local Position Invariance

Here we discuss a simple cyclic thought experiment based on energy conservation following [30], which was used to calculate a quantitative relationship between the violation of the WEP and corresponding violations of gravitational redshift experiments.

Cyclic thought experiment: Two different test bodies X and Y, with corresponding masses m_X and m_Y, respectively, are placed in a uniform gravitational field aligned in the z-direction. Both of them are at rest ($v = 0$), and the test body X is placed at an altitude $z = 0$ and Y at a height h. Imagine now that the test body Y makes a transition to X by emitting a quantum q containing an energy E_q given by

$$E_q = \Delta mc^2 = (m_Y - m_X)c^2, \qquad (5.27)$$

where the nature of the emitted quantum is not specified any further, and it could for example be a photon, a gluon or an α-particle.[4] This quantum q now travels downward in the gravitational field from $z = h$ to $z = 0$ and is absorbed there by the test body X. This absorption leads to a transition from X to Z, so that we end up with two test bodies X and Z at $z = h$ and $z = 0$, respectively, where Z has a corresponding mass m_Z. The test body X at height h now free falls with an acceleration a_X in the negative z-direction. It reaches a velocity v_X at $z = 0$ which can be expressed by

$$v_X^2 = 2a_X h. \qquad (5.28)$$

There it inelastically collides with the test body Z, leaving again two test bodies X and Y. All the leftover energy of this collision is given to the test body Y, which then travels upward in the gravitational potential and experiences a deceleration a_Y.[5] Due to the assumption of energy conservation it must be able to reach a height $z = h$ and therefore needs a velocity v_Y at $z = 0$ given by

---

[4] In Ref. [30] two identical quantum systems in different states were used, instead of two different test bodies. But as the nature of the quantum which is emitted in the transition from Y to X is not restricted, this process can be extended to any two test bodies.

[5] We allow a deviation from the WEP; therefore the two test bodies X and Y may experience a different acceleration a due to their internal structure or composition.

$$v_Y^2 = 2a_Y h. \tag{5.29}$$

Thus one full cycle is fulfilled, and we are again at the starting point of the cyclic thought experiment.

If we take a closer look at the different energy contributions present at the point of collision, we can write

$$\frac{1}{2}m_X v_X^2 + m_X c^2 + m_Z c^2 = m_X c^2 + \frac{1}{2}m_Y v_Y^2 + m_Y c^2, \tag{5.30}$$

where on the left and on the right side the kinetic and rest mass energies before and after the inelastic collision are summed up, respectively. Due to the principle of conservation of energy, these two sides must be equal to each other. This can be rewritten using the expressions for the velocities v_X and v_Y given in Eqs. (5.28) and (5.29), and by canceling out the rest mass energy of X on both sides, as

$$m_X a_X h + m_Z c^2 = m_Y a_Y h + m_Y c^2. \tag{5.31}$$

Our aim is to find a quantitative relation between the violation of the gravitational redshift and the violation of the UFF. First of all we rewrite the equation above as an expression for the mass difference between the test bodies Z and Y which arises due to the gravitational redshift of the quantum q

$$
\begin{aligned}
m_Z - m_Y &= \frac{h}{c^2}(m_Y a_Y - m_X a_X) \\
&= \frac{h}{2c^2}\left[(m_Y - m_X)(a_Y + a_X) + (m_Y + m_X)(a_Y - a_X)\right].
\end{aligned} \tag{5.32}
$$

The gravitational redshift is defined as the shift in frequency divided by the frequency itself (see Eq. 5.20) or stated differently, as the difference between the received and the emitted energy divided by the emitted energy. In our cyclic thought experiment the emitted energy corresponds to the energy of the quantum q at an altitude $z = h$, whereas the received energy is its energy at $z = 0$

$$z = \frac{E_{\text{received}} - E_{\text{emitted}}}{E_{\text{emitted}}} = \frac{(m_Z - m_X)c^2 - (m_Y - m_X)c^2}{(m_Y - m_X)c^2} = \frac{m_Z - m_Y}{m_Y - m_X}. \tag{5.33}$$

The last term can be easily related to the expression given in Eq. (5.32), and we obtain the following connection between the gravitational redshift and the thought experiment

$$z = \frac{h}{2c^2}\left[a_Y + a_X + (a_Y - a_X)\frac{m_Y + m_X}{m_Y - m_X}\right]. \tag{5.34}$$

If we introduce the gravitational acceleration g as the average between the free fall accelerations of the test bodies X and Y, meaning $g = (a_Y + a_X)/2$ we can write

$$z = \frac{gh}{c^2}\left[1 + \frac{a_Y - a_X}{a_Y + a_X}\frac{m_Y + m_X}{m_Y - m_X}\right]. \tag{5.35}$$

By identifying the ratio between the accelerations with the Eötvös ratio as defined in equation (5.9) and allowing deviations from LPI by writing the gravitational redshift as given in equation (5.22) we arrive at

$$(1 + \alpha)\frac{\Delta U}{c^2} = \frac{\Delta U}{c^2}\left[1 + \eta(Y, X)\frac{m_Y + m_X}{2(m_Y - m_X)}\right], \tag{5.36}$$

where the difference in gravitational potential $\Delta U = gh$ is used. From this equation, a simple quantitative relationship between the LPI violation parameter α and the Eötvös ratio η can be deduced

$$\alpha\frac{2(m_X - m_Y)}{m_X + m_Y} = \eta(X, Y). \tag{5.37}$$

If we assume that the two test bodies have about the same mass $m_X \sim m_Y$, we can write the result in the following final form

$$\boxed{\eta(X, Y) = \alpha^A(\zeta_X^A - \zeta_Y^A),} \tag{5.38}$$

where A denotes the form of energy which is assumed to violate LPI and is transferred in the gravitational redshift experiment from which the limit on the parameter α is extracted (e.g., hyperfine energy if a hydrogen maser is used as clock). The fractional energy contribution ζ was already defined in Eq. (5.8).

Using the relation between the Eötvös ratio and the parameter for the strength of the violation of the WEP of interaction A given in Eq. (5.13), we can find an expression for the limit set on the WEP violation parameter by the gravitational redshift experiments

$$|\eta^A| \leq \left|\frac{\zeta_X^B - \zeta_Y^B}{\zeta_X^A - \zeta_Y^A}\right||\alpha^B|, \tag{5.39}$$

which for the simplest case, where the two forms of interactions are identical $A = B$, simplifies to

$$|\eta^A| \leq |\alpha^A|. \tag{5.40}$$

5.4 The Strong Equivalence Principle

In contrast to the EEP, the strong equivalence principle (SEP) includes all types of test bodies, even gravitationally bound ones. Instead of the WEP, it therefore contains the gravitational weak equivalence principle (GWEP) as one of the three pillars, which states that all the test particles behave the same way in a gravitational field, independent of their internal structure or composition. It includes the WEP in the limit where $\sigma \to 0$. We see that the EEP can be treated as a special case of the SEP, when gravitational forces are negligible, and its validation is at the same time a validation of the EEP. Up to now, GR is the only known theory completely fulfilling the SEP.[6] As a consequence its violation would only falsify GR but not other metric theories. The empirical evidence supporting the Einstein equivalence principle supports the conclusion that the only theories of gravity that have a hope of being viable are metric theories. In any metric theory of gravity, matter and non-gravitational fields respond only to the spacetime metric $g_{\mu\nu}$, as is the case of GR. In principle, however, there could exist other gravitational fields besides the metric, such as scalar fields, vector fields and so on. By discussing metric theories it is possible to draw some general conclusions about the nature of gravity in different metric theories, which are reminiscent of the Einstein equivalence principle, but that are subsumed under the name "strong equivalence principle." The strong equivalence principle (SEP) states that:

1. WEP is valid for self-gravitating bodies as well as for test bodies.
2. The outcome of any local test experiment is independent of the velocity of the freely falling apparatus.
3. The outcome at any local test experiment is independent of where and when in the universe it is performed.

The distinction between the SEP and EEP is the inclusion of bodies with self-gravitational interactions, planets, stars and of experiments involving gravitational forces. Empirically it has been found that almost every metric theory other than GR introduces auxiliary gravitational fields, either dynamical or prior geometric, and thus predicts violations of SEP at some level. General relativity seems to be the only viable metric theory that embodies SEP completely.

There have also been experiments to test the WEP with test bodies that have a non-negligible amount of self-gravitational energy, therefore testing the GWEP. The lunar laser ranging yielded $|\eta| \leq 2.1 \times 10^{-13}$ for any possible inequality in the ratios of the gravitational and inertial masses for the Earth and Moon [33]

$$\eta = \left(\frac{m_G}{m_I}\right)_{\text{Earth}} - \left(\frac{m_G}{m_I}\right)_{\text{Moon}} = (-0.8 \pm 1.3) \times 10^{-13}. \tag{5.41}$$

This type of experiments tries to confirm the SEP instead of the EEP.

---

[6] The only other theory known, which fulfills the SEP, is the conformally flat scalar theory developed by Nordström in 1913 [32]. However, it is experimentally ruled out since it predicts no deflection of light.

5.5   Summary

The Einstein equivalence principle is at the hearth of general relativity; therefore finding experimentally stringent tests of the various aspects of the equivalence principle is very important and could, in case of a positive result of a violation, have very profound implications on the knowledge of the fundamental laws of physics. At the same time theoretical studies can lead to better clarification of the consequences of the so far found stringent limits on various alternative theories. We discussed the various aspects of the Einstein equivalence principle, namely the weak equivalence principle, the local Lorentz invariance and the local position invariance. We then elaborated on Schiff's conjecture which states that the three tests of EEP are not independent of each other. Finally, we briefly commented the strong equivalence principle.

References

1. A. Einstein, Grundlagen der allgemeinen Relativitätstheorie. Annalen der Physik **49**, 769–822 (1916)
2. C. M. Will, *Theory and Experiment in Gravitational Physics* (Cambridge University Press, 2018)
3. E.D. Casola, S. Liberati, S. Sonego, Nonequivalence of equivalence principles. Am. J. Phys. **83**, 39–46 (2015)
4. H. Bondi, Negative mass in general relativity. Rev. Mod. Phys. **29**, 423–428 (1957)
5. R.H. Dicke, New research on old gravitation. Science **129**, 3349 (1959)
6. R.H. Dicke, *Mach's Principle and Equivalence*, in *Evidence for Gravitational Theories: Proceedings of Course 20 of the International School of Physics "Enrico Fermi"*, ed. by C. Moller (Academic Press, New York, 1962)
7. C.M. Will, The confrontation between general relativity and experiment. Living Rev. Relativ **17**, 4 (2014)
8. K.S. Thorne, D.L. Lee, A.P. Lightman, Foundations for a theory of gravitation theories. Phys. Rev. D **7**, 3563–3578 (1973)
9. R. von Eötvös, D. Pékar, E. Fekete, Beiträge zum Gesetz der Proportionalität von Trägheit und Gravität. Annalen der Physik **68**, 11–66 (1922)
10. T.A. Wagner, S. Schlamminger, J.H. Gundlach, E.G. Adelberger, Torsion-balance tests of the weak equivalence principle. Class. Quant. Gravity **29**, 184002 (2012)
11. P. Touboul, G. Métris, M. Rodrigues et al., Microscope mission: first results of a space test of the equivalence principle. Phys. Rev. Lett. **119**, 231101 (2017)
12. D.N. Aquilera, H. Ahlers, B. Battelier et al., STE-QUEST-test of the universality of free fall using cold atom interferometry. Class. Quantum Grav. **31**, 115010 (2014)
13. B. Altschul, Q.G. Bailey, L. Blanchet et al., Quantum tests of the Einstein Equivalence Principle with the STE-QUEST space mission. Adv. Space Res. **55**, 501–524 (2015)
14. R.S. Brusa et al., The AEgIS experiment at CERN: measuring antihydrogen free-fall in earths gravitational field to test WEP with antimatter. J. Phys. Conf. Ser. **791**, 012014 (2017)
15. N. Aghanim et al., *Planck 2013 results. XXVII Doppler boosting of the CMB: Eppur si muove*, Astron. & Astrophys. **571**, A27 (2014)
16. A.P. Lightmann, D.L. Lee, Restricted proof that the weak equivalence principle implies the Einstein equivalence principle. Phys. Rew. D **8**, 364–376 (1973)

17. D. Colladay, A. Kostelecký, CPT violation and the standard model. Phys. Rev. D **55**, 6760 (1997)
18. D. Collady, A. Kostelecký, Lorentz-violating extension of the standard model. Phys. Rev. D **58**, 116002 (1998)
19. A. Kostelecký, N. Russel, Data tables for Lorentz and CPT violation. Rev. Mod. Phys. **83**, 11–31 (2011)
20. F. Allmendinger et al., New limit on Lorentz and CPT violating neutron spin interactions using a free precession ^{3}He- ^{129}Xe co-magnetometer. Phys. Rev. Lett. **112**, 110801 (2014)
21. R.V. Pound, G.A. Rebka, Apparent weight of photons. Phys. Rev. Lett. **4**, 337–341 (1960)
22. R.V. Pound, J.L. Snider, Effect of gravity on gamma radiation. Phys. Rev. B **140**, 788–803 (1965)
23. R.F.C. Vessot et al., Test of relativistic gravitation with a space-borne hydrogen maser. Phys. Rev. Lett. **45**, 2081–2084 (1980)
24. P. Delva et al., Test of the gravitational redshift with stable clocks in eccentric orbits: application to Galileo satellites 5 and 6. Class. Quantum Grav. **32**, 232003 (2015)
25. M. Takamoto et al., Test of general relativity by a pair of transportable optical lattice clocks. Nat. Photon. **14**, 411–415 (2020)
26. J.P. Turneaure, C.M. Will, B.F. Farrell, E.M. Mattison, R.F.C. Vessot, Test of the principle of equivalence by a null gravitational red-shift experiment. Phys. Rev. D **27**, 1705–1714 (1983)
27. J. Guéna et al., Improved tests of local position invariance using ^{87}Rb and ^{133}Cs fountains. Phys. Rev. Lett. **109**, 80801 (2012)
28. L.I. Schiff, On experimental tests of the general theory of relativity. Am. J. Phys. **28**, 340–343 (1960)
29. W.T. Ni, *Equivalence principles and electromagnetism*. Phys. Rev. Lett. **38**, 301–304 (1977)
30. K. Nordtvedt, Quantitative relationship between clock gravitational "red-shiftâŁž violations and nonuniversality of free-fall rates in nonmetric theories of gravity. Phys. Rev. D **11**, 245–247 (1975)
31. M.P. Haugan, Energy conservation and the principle of equivalence. Ann. Phys. **118**, 156–186 (1979)
32. G. Nordström, Zur Theorie der Gravitation vom Standpunkt des Relativitätsprinzips. Annalen der Physik **347**, 533–554 (1913)
33. J.G. Williams, S.G. Turyshev, D.H. Boggs, Lunar laser ranging tests of the equivalence principle. Class. Quant. Gravity **29**, 184004 (2012)

Solutions

In the following we give the solutions of the problems listed at the end of the various chapters.

Solutions to Problems in Chap. 2

Problem 2.1 *(Metric of a static star)*

(a) We start from the general static, spherical symmetric metric

$$ds^2 = \exp[2\alpha(r)]dt^2 - \exp[2\beta(r)]dr^2 - r^2 d\Omega^2. \tag{S1}$$

From the normalization of the velocity in the fluid rest frame we obtain $u_t = \sqrt{g_{tt}}$, which then straightforwardly leads to the energy-momentum tensor

$$T_{\mu\nu} = \mathrm{diag}\left\{\exp[2\alpha]\,\rho,\ \exp[2\beta]\,p,\ r^2 p,\ r^2 \sin^2(\theta)\,p\right\}. \tag{S2}$$

(b) The Ricci scalar for the above metric ansatz reads as

$$R = 2\exp[-2\beta]\left(\partial_r^2\alpha + (\partial_r\alpha)^2 - \partial_r\alpha\partial_r\beta + \frac{2}{r}(\partial_r\alpha - \partial_r\beta) + \frac{1}{r^2}(1 - \exp[2\beta])\right). \tag{S3}$$

For the time component of the Einstein equations we thus have

$$R_{tt} - \frac{1}{2}Rg_{tt} = 8\pi\,G\,T_{tt}\ , \tag{S4}$$

$$-\exp[2(\alpha - \beta)]\left(-\frac{2}{r}\partial_r\beta + \frac{1}{r^2}(1 - \exp[2\beta])\right) = 8\pi\,G\rho\exp[2\alpha]\ , \tag{S5}$$

$$\frac{1}{r^2}\partial_r m(r) = 4\pi\rho\ , \tag{S6}$$

P. Jetzer, *Applications of General Relativity*, UNITEXT for Physics, https://doi.org/10.1007/978-3-030-95718-6

163

where $m(r)$ is defined in Eq. (2.185). This differential equation can be integrated to obtain the mass within a given radius r

$$m(r) = 4\pi \int_0^r \rho(r')r'^2 dr'. \tag{S7}$$

Note that it is different from the integrated energy density, which is obtained integrating the density over the volume element $\sqrt{\gamma}d^3x$. Here γ is the determinant of the spatial metric. In contrast to the above, the integrated energy density accounts for the gravitational binding energy as well.

With the replacement of $m(r)$ we can immediately write down the radial metric element

$$\exp[2\beta] = \left(1 - \frac{2Gm(r)}{r}\right)^{-1}, \tag{S8}$$

which has remarkable similarity with the metric element of the Schwarzschild solution. For the radial component we have

$$R_{rr} - \frac{1}{2}Rg_{rr} = 8\pi G T_{rr}, \tag{S9}$$

$$\frac{2}{r}\partial_r\alpha - \frac{2G}{r^3}\exp[2\beta]m(r) = 8\pi Gp, \tag{S10}$$

thus

$$\frac{d\alpha}{dr} = \frac{4\pi Gr^3 p + Gm(r)}{r[r - 2Gm(r)]}. \tag{S11}$$

(c) The Bianchi identity or energy-momentum conservation reads as

$$\nabla_\mu T^{\mu\nu} = \partial_\mu T^{\mu\nu} + \Gamma^\mu_{\mu\rho}T^{\rho\nu} + \Gamma^\nu_{\mu\rho}T^{\mu\rho} = 0. \tag{S12}$$

While the $\nu = t$ equation is trivial, the $\nu = r$ equation leads to

$$\begin{aligned}
\nabla_\mu T^{\mu r} &= \partial_r(\exp[-2\beta]p) + T^{rr}(\Gamma^r_{rr} + \Gamma^\theta_{\theta r} + \Gamma^\phi_{\phi r} + \Gamma^t_{tr}) \\
&\quad + (\Gamma^r_{tt}T^{tt} + \Gamma^r_{rr}T^{rr} + \Gamma^r_{\phi\phi}T^{\phi\phi} + \Gamma^r_{\theta\theta}T^{\theta\theta}) \\
&= \exp[-2\beta]\{(\partial_r p - 2p\partial_r\beta) \\
&\quad + p(\partial_r\beta + 2/r + \partial_r\alpha) + (\rho\partial_r\alpha + \partial_r\beta p - 2p/r)\} \\
&= \exp[-2\beta]\{(\rho + p)\partial_r\alpha + \partial_r p\} \stackrel{!}{=} 0.
\end{aligned} \tag{S13, S14}$$

From this we obtain

$$(\rho + p)\frac{d\alpha}{dr} = -\frac{dp}{dr}. \tag{S15}$$

Plugging in Eq. (S11) we finally have

$$\frac{dp}{dr} = -\frac{(\rho + p)\left[Gm(r) + 4\pi Gr^3 p\right]}{r\left[r - 2Gm(r)\right]}. \tag{S16}$$

(d) Now we assume $\rho = \rho_* = $ const. out to the surface of the star yielding

$$m(r) = \frac{4\pi}{3}\rho_* r^3 = M\frac{r^3}{R^3}, \tag{S17}$$

where M is the total mass of the star and from now on R will denote the radius of the star. From Eq. (S16) we obtain by separation of variables

$$\frac{dp}{p^2 + \frac{4}{3}\rho_* p + \frac{\rho_*^2}{3}} = \frac{4\pi\,Grdr}{\frac{8\pi G}{3}r^2\rho_* - 1}, \tag{S18}$$

$$\frac{dp}{\left(p + \frac{2}{3}\rho_*\right)^2 - \frac{\rho_*^2}{9}} = \frac{3}{4\rho_*}\frac{\frac{16\pi G}{3}r\rho_* dr}{\frac{8\pi G}{3}r^2\rho_* - 1}. \tag{S19}$$

Using that the pressure vanishes at the surface of the star we can integrate to obtain

$$-\frac{1}{2}\frac{3}{\rho_*}\ln\left[\frac{p' + \frac{2}{3}\rho_* + \frac{1}{3}\rho_*}{p' + \frac{2}{3}\rho_* - \frac{1}{3}\rho_*}\right]_{p'=p}^{0} = \frac{3}{4\rho_*}\ln\left[\frac{8\pi G}{3}r'^2\rho_* - 1\right]_{r'=r}^{R} \tag{S20}$$

$$\ln\left[\frac{p + \rho_*}{3p + \rho_*}\right] = \frac{1}{2}\ln\left[\frac{2GMR^2 - R^3}{2GMr^2 - R^3}\right]. \tag{S21}$$

Solving for $p(r)$ yields the radial pressure distribution

$$p(r) = \rho_*\frac{\sqrt{R^3 - 2GMR^2} - \sqrt{R^3 - 2GMr^2}}{\sqrt{R^3 - 2GMr^2} - 3\sqrt{R^3 - 2GMR^2}}. \tag{S22}$$

For the star with maximal mass the pressure diverges at $r = 0$ and there is no static solution, i.e., the star collapses. The maximum mass corresponds to the case where the denominator vanishes at $r = 0$

$$R = 9(R - 2GM), \tag{S23}$$

leading to

$$M < \frac{4}{9}\frac{R}{G}. \tag{S24}$$

This result remains true for more general equations of state and is known as Buchdahl's theorem.

Now it remains to calculate the missing metric coefficients. From Eq. (S11) we have

$$\frac{d\alpha}{dr} = -\frac{d\ln(p + \rho_*)}{dr}. \tag{S25}$$

The boundary conditions can be set up at the surface of the star, where the interior solution goes over into the exterior Schwarzschild solution, thus $\rho + p|_{r=R} = \rho_*$ and

$$\exp\left[\alpha(R)\right] = \sqrt{1 - \frac{2GM}{R}}. \tag{S26}$$

In summary, we have for the metric elements

$$\exp\left[\alpha\right] = \frac{3}{2}\sqrt{1 - \frac{2GM}{R}} - \frac{1}{2}\sqrt{1 - \frac{2GMr^2}{R^3}}, \tag{S27}$$

$$\exp\left[-\beta\right] = \sqrt{1 - \frac{2Gm(r)}{r}} = \sqrt{1 - \frac{2GMr^2}{R^3}}. \tag{S28}$$

Problem 2.2 *(Linearized field equations)*

(a) (i) We show that the condition $g_{\mu\rho}g^{\rho\nu} = \delta_\mu^\nu$ is satisfied:

$$g_{\mu\rho}g^{\rho\nu} = (\eta_{\mu\rho} + h_{\mu\rho})(\eta^{\rho\nu} - h^{\rho\nu}) = \eta_{\mu\rho}\eta^{\rho\nu} + \eta^{\rho\nu}h_{\mu\rho} - \eta_{\mu\rho}h^{\rho\nu}$$
$$= \delta_\mu^\nu + h_\mu^{\ \nu} - h_\mu^{\ \nu} = \delta_\mu^\nu. \tag{S29}$$

Then it follows:

$$h = h^\mu_{\ \mu} = g^{\mu\nu}h_{\mu\nu} = (\eta^{\mu\nu} - h^{\mu\nu})h_{\mu\nu} = \eta^{\mu\nu}h_{\mu\nu}. \tag{S30}$$

Thus, indices of tensors that are of first order in $h_{\mu\nu}$ can be raised and lowered using the Minkowski metric.

(ii) To first order in $h_{\mu\nu}$ the Christoffel symbols are

$$\Gamma^\rho_{\mu\nu} = \frac{1}{2}\eta^{\rho\lambda}\left(\partial_\mu h_{\nu\lambda} + \partial_\nu h_{\lambda\mu} - \partial_\lambda h_{\mu\nu}\right). \tag{S31}$$

The $\Gamma\Gamma$ terms in the Ricci tensor contribute to higher orders only, s.t.

$$R_{\mu\nu} = \partial_\rho \Gamma^\rho_{\nu\mu} - \partial_\nu \Gamma^\rho_{\rho\mu}$$

$$= \frac{1}{2}\eta^{\rho\lambda}\left(\partial_\rho\partial_\nu h_{\mu\lambda} + \partial_\rho\partial_\mu h_{\lambda\nu} - \partial_\rho\partial_\lambda h_{\nu\mu}\right)$$

$$- \frac{1}{2}\eta^{\rho\lambda}\left(\partial_\nu\partial_\rho h_{\mu\lambda} + \partial_\nu\partial_\mu h_{\lambda\rho} - \partial_\nu\partial_\lambda h_{\rho\mu}\right)$$

$$= \frac{1}{2}\eta^{\rho\lambda}\left(\partial_\rho\partial_\mu h_{\lambda\nu} - \partial_\nu\partial_\mu h_{\lambda\rho} - \partial_\rho\partial_\lambda h_{\nu\mu} + \partial_\nu\partial_\lambda h_{\rho\mu}\right)$$

$$= -\frac{1}{2}\left(\Box h_{\mu\nu} + h_{,\mu,\nu} - h^\rho{}_{\mu,\rho,\nu} - h^\rho{}_{\nu,\rho,\mu}\right).$$

(iii) Trivial.

(b) (i) Use that $\frac{\partial \tilde{x}^\mu}{\partial x^\rho} = \delta^\mu_\rho - \partial_\rho \xi^\mu$ and thus $\frac{\partial x^\rho}{\partial \tilde{x}^\mu} = \delta^\rho_\mu + \partial^\rho \xi_\mu$. The metric transforms as

$$\tilde{g}_{\mu\nu} = \frac{\partial x^\rho}{\partial \tilde{x}^\mu}\frac{\partial x^\lambda}{\partial \tilde{x}^\nu}g_{\rho\lambda}$$
$$= \left(\delta^\rho_\mu + \partial^\rho \xi_\mu\right)\left(\delta^\lambda_\nu + \partial^\lambda \xi_\nu\right)\left(\eta_{\rho\lambda} + h_{\rho\lambda}\right) \qquad (S32)$$
$$= \eta_{\mu\nu} + h_{\mu\nu} + \left(\partial_\mu \xi_\nu + \partial_\nu \xi_\mu\right).$$

Identifying with $\tilde{g}_{\mu\nu} = \eta_{\mu\nu} + \tilde{h}_{\mu\nu}$, we find

$$\tilde{h}_{\mu\nu} = h_{\mu\nu} + \left(\partial_\mu \xi_\nu + \partial_\nu \xi_\mu\right).$$

(ii) We obtain the result by seeing that (in this gauge) the LHS of the linearized field equations becomes

$$\Box h_{\mu\nu} + h_{,\mu,\nu} - \frac{1}{2}h_{,\mu,\nu} - \frac{1}{2}h_{,\nu,\mu} = \Box h_{\mu\nu}. \qquad (S33)$$

iii) For $T_{\mu\nu} = \text{diag}\{\rho, 0, 0, 0\}$, we have $T = \rho$ and then:

$$\Box h_{\mu\nu} = 8\pi G\rho\, \delta_{\mu\nu}. \qquad (S34)$$

Since the source is static, ρ does not depend on t, and thus $h_{\mu\nu}$ does not depend on t and $\Box$ reduces to $\triangle$. We finally have:

$$\triangle h_{\mu\nu} = 8\pi G\rho\, \delta_{\mu\nu}, \quad \Rightarrow \quad \triangle\left(\frac{h_{\mu\nu}}{2}\right) = 4\pi G\rho\, \delta_{\mu\nu}, \qquad (S35)$$

which is the Poisson equation with

$$h_{\mu\nu} = 2\phi\delta_{\mu\nu}, \qquad (S36)$$

where ϕ is the Newtonian potential associated with ρ. Finally, we obtain the metric:

$$ds^2 = g_{\mu\nu}dx^\mu dx^\nu = (1 + 2\phi)dt^2 - (1 - 2\phi)\left(dx^2 + dy^2 + dz^2\right). \quad \text{(S37)}$$

(c) Using $h_{\mu\nu} = \gamma_{\mu\nu} + \frac{1}{2}\eta_{\mu\nu}h$, the Lorentz gauge implies

$$h_{\mu\nu}{}^{,\nu} = \gamma_{\mu\nu}{}^{,\nu} + \frac{1}{2}\eta_{\mu\nu}h^{,\nu} = \frac{1}{2}\eta_{\mu\nu}h^{,\nu} = \frac{1}{2}h_{,\mu}. \quad \text{(S38)}$$

Then, the Ricci tensor is

$$R_{\mu\nu} = -\frac{1}{2}\left(\Box h_{\mu\nu} + h_{,\mu,\nu} - \frac{1}{2}h_{,\mu,\nu} - \frac{1}{2}h_{,\nu,\mu}\right) = -\frac{1}{2}\Box h_{\mu\nu} \quad \text{(S39)}$$

and the Ricci scalar

$$R = \eta^{\mu\nu}R_{\mu\nu} = -\frac{1}{2}\Box h. \quad \text{(S40)}$$

The Einstein tensor is thus

$$G_{\mu\nu} = R_{\mu\nu} - \frac{1}{2}\eta_{\mu\nu}R = -\frac{1}{2}\left(\Box h_{\mu\nu} - \frac{1}{2}\eta_{\mu\nu}\Box h\right)$$
$$= -\frac{1}{2}\left(\Box\gamma_{\mu\nu} + \frac{1}{2}\eta_{\mu\nu}\Box h - \frac{1}{2}\eta_{\mu\nu}\Box h\right) = -\frac{1}{2}\Box\gamma_{\mu\nu}. \quad \text{(S41)}$$

Then, the Einstein equations $G_{\mu\nu} = 8\pi G T_{\mu\nu}$ become

$$\Box\gamma_{\mu\nu} = -16\pi G T_{\mu\nu}. \quad \text{(S42)}$$

Problem 2.3 *(Out of plane precession of S2 orbit)*

The time derivative of the orbital momentum $\vec{l} = \vec{r} \wedge \vec{v}$ is

$$\frac{d\vec{l}}{dt} = \vec{r} \wedge \frac{d\vec{v}}{dt},$$
$$= -\vec{r} \wedge \nabla\phi + 2\vec{r} \wedge \left(\vec{\Omega} \wedge \vec{v}\right),$$
$$= 2\vec{r} \wedge \left(\vec{\Omega} \wedge \vec{v}\right). \quad \text{(S43)}$$

We now assume $\vec{l} \perp \vec{S}$ and circular orbits. A good choice for the radius of the circular orbit may be the actual mean distance between focal point and S2 as obtained by integrating over the true anomaly.[1] This results in

$$\bar{r} = \frac{a_r(1-e^2)}{2\pi} \int\limits_0^{2\pi} d\phi \frac{1}{1+e\cos\phi} = a_r\sqrt{1-e^2} = 2.2 \times 10^{-3} \text{ pc}, \qquad (S44)$$

where a_r is the semimajor axis and e is the ellipticity.
The orbital time for S2 is given by

$$T_o \approx 2\pi\sqrt{\frac{a_r^3}{GM}} = 14.8\text{year}. \qquad (S45)$$

It is reasonable to assume that the frequency of the orbital plane precession is much smaller than the frequency of the orbital motion of S2 (this can be checked a posteriori). Under this assumption we can calculate the precession integrating over the unperturbed orbit. With $\vec{S} = S\vec{e}_y$, $\vec{r} = \bar{r}\left[\cos(\varphi)\,\vec{e}_x + \sin(\varphi)\,\vec{e}_y\right]$ and $\vec{v} = \bar{r}\omega\left[-\sin(\varphi)\vec{e}_x + \cos(\varphi)\vec{e}_y\right]$ we obtain

$$\frac{d\vec{l}}{dt} = \frac{4GS\omega}{c^2\bar{r}}\left(\sin^2(\varphi)\vec{e}_x - \sin(\varphi)\cos(\varphi)\vec{e}_y\right). \qquad (S46)$$

After one orbit the change in angular momentum is

$$\Delta\vec{l} = \int \frac{d\vec{l}}{dt}dt = \frac{T_o}{2\pi}\int \frac{d\vec{l}}{dt}d\varphi = 4\pi\frac{GS}{c^2\bar{r}}\vec{e}_x. \qquad (S47)$$

The total change in the orbital angular momentum is perpendicular to the angular momentum, so the gravitomagnetic field does not change the magnitude of the angular momentum vector, but just its direction. The maximal angular frequency of the precession can now be estimated as

$$\Omega_{\max} = \frac{\Delta l}{lT_o} = 2\frac{G^2M^2}{c^3\bar{r}^3}a, \qquad (S48)$$

where $a \in [0, 1]$ is the spin parameter of the central black hole. A comparison with the orbital frequency $\Omega_o = \frac{2\pi}{T_o}$ gives, setting $a = 1$,

$$\frac{\Omega_{\max}}{\Omega_o} \sim 10^{-6} \ll 1, \qquad (S49)$$

which is in agreement with our a priori assumption. Estimating the "real" out-of-plane precession as $\Omega_{\text{real}} = \frac{75}{360}\frac{2\pi}{t_{S2}}$, we obtain the constraint on the black hole spin

---

[1] The true anomaly is the angle between the direction of periapsis and the current position of the body, as seen from the main focus of the ellipse.

parameter:

$$a \geq \left(4\pi \frac{G^2M^2}{c^3\bar{r}^3}\right)^{-1} \Omega_{\text{real}} \approx 0.03. \tag{S50}$$

Since the lower bound lies within the interval allowed by general relativity, the estimation we have performed is not able to exclude "gravitomagnetic" precession as a source of S2 inclination. Note that the spin parameter of the black hole in the center of our galaxy is considered to be much higher than this lower bound ($a \sim 0.6$). Remember, however, that the approximations we have done are rather drastic, and we are not able, with these simple calculations, to say how realistic the hypothesis of gravitomagnetic precession is.

Problem 2.4 *(Gravitational field of a moving particle)*

We are considering a particle of mass M moving with constant velocity $\vec{V}$. In the particle rest frame Σ' we have

$$h'_{\mu\mu}(\vec{r}) = -\frac{2G}{c^2} \int d^3r' \frac{\rho(r')}{|\vec{r} - \vec{r}'|} = -\frac{2GM}{c^2} \frac{1}{|\vec{r} - \vec{r}_M|} \tag{S51}$$

and $h'_{0i} = 0$. Here $\vec{r}_M$ is the position of the mass M. If we now transform to a general frame Σ, the metric changes according to $g_{\mu\nu} = \alpha^\kappa_\mu \alpha^\rho_\nu g'_{\kappa\rho}$. Since we are working at linear order in h and the mass M is moving on a straight line, the transformation is nothing but a Lorentz transformation $\alpha^\mu_\nu = \Lambda^\mu_\nu + \mathcal{O}(h)$. For the transformation matrix we have up

$$\Lambda^\mu_\nu = \begin{pmatrix} 1 & V^i/c \\ V^i/c & \delta_{ij} \end{pmatrix}, \tag{S52}$$

where we neglected terms $\mathcal{O}(V^2/c^2)$, i.e. set $\gamma = 1$. Note that we only need to transform the perturbation since $\Lambda^\kappa_\mu \Lambda^\rho_\nu \eta_{\kappa\rho} = \eta_{\mu\nu}$ and are thus left with

$$h_{\mu\nu} = \Lambda^\kappa_\mu \Lambda^\rho_\nu h'_{\kappa\rho} . \tag{S53}$$

Evaluating the above expression we obtain for the metric perturbation in Σ

$$h_{\mu\mu} = h'_{\mu\mu} + \mathcal{O}\left(V^2/c^2\right) , \qquad h_{0i} = 2\frac{V_i}{c}h'_{00} + \mathcal{O}\left(V^2/c^2\right) . \tag{S54}$$

The mass M is moving on a straight line in Σ; thus its position as a function of time can be written as

$$\vec{r}_M(t) = \vec{r}_{M,0} + \vec{V}t. \tag{S55}$$

The gravitomagnetic potential is

$$\vec{h}(\vec{r}) = h_{0i}(\vec{r}) = -\frac{4GM}{c^3} \frac{\vec{V}}{|\vec{r} - \vec{r}_M(t)|}, \tag{S56}$$

For the gravitomagnetic field we thus have

$$\vec{\Omega}(\vec{r}) = -\frac{c}{2}\vec{\nabla}_{\vec{r}} \times \vec{h}(\vec{r}) = \frac{2GM}{c^2} \frac{\vec{V} \times (\vec{r} - \vec{r}_M(t))}{|\vec{r} - \vec{r}_M(t)|^3}. \tag{S57}$$

The equation of motion is calculated in analogy to the derivation presented in Sect. 2.6.2. The velocity of the test mass m will be denoted by a lowercase $\vec{v}$. Evaluating the geodesic equation we obtain

$$\frac{dv^i}{dt} = -\Gamma^i_{\alpha\beta}u^\alpha u^\beta = -\Gamma^i_{00}u^0 u^0 - 2\Gamma^i_{0j}u^0 u^j, \tag{S58}$$

$$= -c^2\Gamma^i_{00} - 2\Gamma^i_{0j}cv^j_m, \tag{S59}$$

where we have neglected terms $\mathcal{O}(v^2/c^2)$. For the Christoffel symbols we get

$$\Gamma^i_{00} = \frac{1}{2}\eta^{ij}\left(\partial_0 h_{j0} + \partial_0 h_{0j} - \partial_j h_{00}\right) = \partial_0 h^i_0 - \frac{1}{2}\partial^i h_{00}, \tag{S60}$$

$$\Gamma^i_{0j} = \frac{1}{2}\eta^{ik}\left(\partial_0 h_{jk} + \partial_j h_{0k} - \partial_k h_{0j}\right) = \frac{1}{2}\partial_0 h^i_j + \frac{1}{2}\eta^{ik}\left(\partial_j h_{0k} - \partial_k h_{0j}\right). \tag{S61}$$

In contrast to the results in Sect. 2.6, we now have a time-dependent metric perturbation. The time dependence arises via $\vec{r}_M(t)$, such that the time derivative can be rewritten as

$$\frac{\partial h_{\mu\nu}}{\partial t} = -\vec{V} \cdot \vec{\nabla}_{\vec{r}} h_{\mu\nu}. \tag{S62}$$

Hence we have for the equation of motion of the test mass m

$$\frac{dv_i}{dt} = -c^2\partial_0 h_{i0} + \frac{c^2}{2}\partial_i h_{00} - \frac{c}{2}v^j\partial_0 h_{ij} - \frac{c}{2}v^j\left(\partial_j h_{0i} - \partial_i h_{0j}\right), \tag{S63}$$

$$= -c^2\partial_0 h_{i0} + \frac{c^2}{2}\partial_i h_{00} - \frac{c}{2}v^j\partial_0 h_{ij} + \epsilon_{ikj}\Omega^k v^j. \tag{S64}$$

Neglecting $\partial_0 h_{i0} = \mathcal{O}(V^2/c^2)$ we finally have

$$\frac{d\vec{v}}{dt} = -GM\frac{(\vec{r} - \vec{r}_M)\cdot\vec{V}}{|\vec{r} - \vec{r}_M|^3}\vec{v} - GM\frac{\vec{r} - \vec{r}_M}{|\vec{r} - \vec{r}_M|^3} + \vec{\Omega} \times \vec{v}. \tag{S65}$$

We see that the gravitomagnetic forces due to moving masses are of order $\mathcal{O}(vV/c^2)$.

Solutions to Problems in Chap. 3

Problem 3.1 *(Particles in the field of a gravitational wave)*

We start from the parametric form of an ellipse centered on the origin

$$r(\varphi) = \frac{b}{\sqrt{1 - e^2 \cos^2(\varphi)}}. \tag{S66}$$

Thus we have for $e^2 \ll 1$

$$r^2(\varphi) = b^2 \left(1 + \frac{e^2}{2} (1 + \cos(2\varphi)) \right). \tag{S67}$$

Our goal is to bring this in the form of the distortion of a circle due to an incoming gravitational wave

$$r^2(\varphi) = R^2 \left(1 - 2h \cos(\omega t) \cos(2\varphi) \right), \tag{S68}$$

which can be achieved by setting

$$R^2 = b^2 \left(1 + \frac{e^2}{2} \right) = \text{const.}, \qquad\qquad e^2 = -4h \cos(\omega t). \tag{S69}$$

Using $b^2 = R^2(1 - e^2/2)$ and $a^2 - b^2 = a^2 e^2$ we obtain

$$a = R(1 - h \cos(\omega t)), \qquad\qquad b = R(1 + h \cos(\omega t)). \tag{S70}$$

Equations (S69) and (S70) describe the time dependence of the ellipticity and the semiminor and semimajor axis. It might seem troublesome that for $-\pi/2 \le \omega t \le \pi/2$ we have $b > a$ and $e^2 < 0$. But this can be interpreted as a phase shift by $\pi/2$ which means that the ellipse is rotated by 90° with respect to its standard orientation.

For the second polarization we have to add a phase factor of $-\pi/4$ to the ellipse Eq. (S66).

Problem 3.2 *(Non-monochromatic gravitational waves)*

We start from Eq. (3.54):

$$
\begin{aligned}
R^{(2)}_{\mu\kappa} = {} & -\frac{h^{\lambda\nu}}{2} \left[\frac{\partial^2 h_{\lambda\nu}}{\partial x^\mu \partial x^\kappa} + \frac{\partial^2 h_{\mu\kappa}}{\partial x^\lambda \partial x^\nu} - \frac{\partial^2 h_{\mu\nu}}{\partial x^\lambda \partial x^\kappa} - \frac{\partial^2 h_{\lambda\kappa}}{\partial x^\mu \partial x^\nu} \right] \\
& + \frac{1}{4} \left[\frac{\partial h^\nu{}_\sigma}{\partial x^\nu} + \frac{\partial h^\nu{}_\sigma}{\partial x^\nu} - \frac{\partial h^\nu{}_\nu}{\partial x^\sigma} \right] \left[\frac{\partial h^\sigma{}_\mu}{\partial x^\kappa} + \frac{\partial h^\sigma{}_\kappa}{\partial x^\mu} - \frac{\partial h_{\mu\kappa}}{\partial x_\sigma} \right] \\
& - \frac{1}{4} \left[\frac{\partial h_{\sigma\kappa}}{\partial x^\lambda} + \frac{\partial h_{\sigma\lambda}}{\partial x^\kappa} - \frac{\partial h_{\lambda\kappa}}{\partial x^\sigma} \right] \left[\frac{\partial h^\sigma{}_\mu}{\partial x_\lambda} + \frac{\partial h^{\sigma\lambda}}{\partial x^\mu} - \frac{\partial h^\lambda{}_\mu}{\partial x_\sigma} \right]. \tag{S71}
\end{aligned}
$$

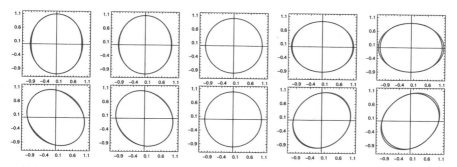

Fig. S.1 Time dependence of the elliptical distortion for $\omega t = 0, \pi/4, \pi/2, 3\pi/4, \pi$ and $h = 0.1$. The blue curve shows the ellipse with the approximation $e \ll 1$, and the red curve shows the exact solution. *Top row:* h_+ polarization *Bottom row:* $h_\times$ polarization

We will now pick a gauge in which $h_{\alpha\beta}$ is transverse $\partial_\mu h^{\mu\nu} = 0$ and traceless $h^\mu_\mu = 0$. Thus the trace reverses metric and the metric agree $h_{\mu\nu} = \bar{h}_{\mu\nu}$, and we will use h from now on. In this gauge the first bracket in the second line vanishes automatically. For the remaining terms we have to perform partial integrations neglecting the surface terms

$$h^{\lambda\nu}\left[\frac{\partial^2 h_{\lambda\nu}}{\partial x^\mu \partial x^\kappa} + \frac{\partial^2 h_{\mu\kappa}}{\partial x^\lambda \partial x^\nu} - \frac{\partial^2 h_{\mu\nu}}{\partial x^\lambda \partial x^\kappa} - \frac{\partial^2 h_{\lambda\kappa}}{\partial x^\mu \partial x^\nu}\right]$$

$$= -\frac{\partial h^{\lambda\nu}}{\partial x^\mu}\frac{\partial h_{\lambda\nu}}{\partial x^\kappa} - \frac{\partial h^{\lambda\nu}}{\partial x^\lambda}\frac{\partial h_{\mu\kappa}}{\partial x^\nu} + \frac{\partial h^{\lambda\nu}}{\partial x^\lambda}\frac{\partial h_{\mu\nu}}{\partial x^\kappa} + \frac{\partial h^{\lambda\nu}}{\partial x^\nu}\frac{\partial h_{\lambda\kappa}}{\partial x^\mu}$$

$$= -\partial_\mu h^{\lambda\nu}\partial_\kappa h_{\lambda\nu}\,. \tag{S72}$$

Except for the first one, all the terms vanish because h is transverse. It remains to work out the third line in Eq. (S71), again integrating by parts and neglecting boundary terms

$$\left[\frac{\partial h_{\sigma\kappa}}{\partial x^\lambda} + \frac{\partial h_{\sigma\lambda}}{\partial x^\kappa} - \frac{\partial h_{\lambda\kappa}}{\partial x^\sigma}\right]\left[\frac{\partial h^\sigma_\mu}{\partial x_\lambda} + \frac{\partial h^{\sigma\lambda}}{\partial x^\mu} - \frac{\partial h^\lambda_\mu}{\partial x_\sigma}\right]$$

$$= -h_{\sigma\kappa}\partial_\lambda\partial^\lambda h^\sigma_\mu - h_{\sigma\kappa}\partial_\mu\partial_\lambda h^{\sigma\lambda} + h_{\sigma\kappa}\partial^\sigma\partial_\lambda h^\lambda_\mu$$

$$- \partial_\kappa\partial^\lambda h_{\sigma\lambda}h^\sigma_\mu + \partial_\kappa h_{\sigma\lambda}\partial_\mu h^{\sigma\lambda} + \partial_\kappa\partial^\sigma h_{\sigma\lambda}h^\lambda_\mu$$

$$+ h_{\lambda\kappa}\partial^\lambda\partial_\sigma h^\sigma_\mu + h_{\lambda\kappa}\partial_\mu\partial\sigma h^{\sigma\lambda} - h_{\lambda\kappa}\partial_\sigma\partial^\sigma h^\lambda_\mu$$

$$= \partial_\kappa h_{\sigma\lambda}\partial_\mu h^{\sigma\lambda}\,. \tag{S73}$$

With the wave equation $\Box h_{\mu\nu} = 0$ and the transversality condition one can cancel all the terms except for the second term in the second line.

$$\langle R^{(2)}_{\mu\kappa}\rangle = \frac{1}{4}\partial_\mu h^{\lambda\nu}\partial_\kappa h_{\lambda\nu}\,. \tag{S74}$$

Note that

$$\langle R^{(2)} \rangle = \frac{1}{4} \langle \partial^\kappa h^{\lambda\nu} \partial_\kappa h_{\lambda\nu} \rangle = -\frac{1}{4} \langle h^{\lambda\nu} \partial^\kappa \partial_\kappa h_{\lambda\nu} \rangle = 0 , \tag{S75}$$

due to the wave equation. From Eq. (3.49) we have

$$t_{\text{grav}}^{\mu\nu} = \frac{c^4}{16\pi G} \left[2R_{\mu\nu}^{(2)} - \eta_{\mu\nu} \eta^{\rho\sigma} R_{\rho\sigma}^{(2)} \right] . \tag{S76}$$

Thus

$$t_{\mu\nu}^{\text{grav}} = \frac{c^4}{32\pi G} \langle \partial_\mu h^{\alpha\beta} \partial_\nu h_{\alpha\beta} \rangle . \tag{S77}$$

Using the three remaining gauge modes we can set $h^{i0} = 0$. From the transversality condition we have for the time component

$$\partial_0 h^{00} = \partial_i h^{i0} = 0. \tag{S78}$$

For gravitational waves we are interested in the dynamical degrees of freedom and not in constant metric perturbations corresponding to the Newtonian potential, and thus we can set $h^{00} = 0$ and are left with a purely spatial metric perturbation

$$t_{\mu\nu}^{\text{grav}} = \frac{c^4}{32\pi G} \langle \partial_\mu h^{ij} \partial_\nu h_{ij} \rangle. \tag{S79}$$

The time–time component reads as

$$t_{00}^{\text{grav}} = \frac{c^4}{32\pi G} \langle \partial_0 h^{ij} \partial_0 h_{ij} \rangle , \tag{S80}$$

and for plane wave perturbations $h_{\mu\nu} = e_{\mu\nu} \exp\left[-ik_\lambda x^\lambda\right] + \text{c.c.}$ it simplifies to

$$t_{00}^{\text{grav}} = \frac{c^4 \omega^2}{16\pi G} e^{ij} e_{ij}. \tag{S81}$$

Problem 3.3 (*The quadrupole radiation formula*)

(a) We start with

$$\gamma_{\mu\nu}(t, \boldsymbol{x}) = 4G \int d^3 y \frac{1}{|\boldsymbol{x} - \boldsymbol{y}|} T_{\mu\nu} \left(t - |\boldsymbol{x} - \boldsymbol{y}|, \boldsymbol{y}\right) . \tag{S82}$$

We take the Fourier transform of this

$$
\begin{aligned}
\widetilde{\gamma}_{\mu\nu}(\omega, \boldsymbol{x}) &= \frac{1}{\sqrt{2\pi}} \int dt\, e^{-i\omega t} \gamma_{\mu\nu}(t, \boldsymbol{x}) \\
&= \frac{4G}{\sqrt{2\pi}} \int dt\, e^{-i\omega t} \int d^3 y \frac{1}{|\boldsymbol{x}-\boldsymbol{y}|} T_{\mu\nu}\left(t - |\boldsymbol{x}-\boldsymbol{y}|, \boldsymbol{y}\right) \\
&= 4G \int d^3 y \frac{e^{-i\omega|\boldsymbol{x}-\boldsymbol{y}|}}{|\boldsymbol{x}-\boldsymbol{y}|} \frac{1}{\sqrt{2\pi}} \int dt_r e^{-i\omega t_r} T_{\mu\nu}\left(t_r, \boldsymbol{y}\right) \\
&= 4G \int d^3 y \frac{e^{-i\omega|\boldsymbol{x}-\boldsymbol{y}|}}{|\boldsymbol{x}-\boldsymbol{y}|} \widetilde{T}_{\mu\nu}(\omega, \boldsymbol{y}) ,
\end{aligned}
\tag{S83}
$$

where in the first step we replaced t with t_r and noted that $dt = dt_r$. The last step is then just definition of the Fourier transform of $T_{\mu\nu}$.

(b) Far away from the source we set $|\boldsymbol{x}-\boldsymbol{y}| = r$. We also only focus on the spatial components now, as the time components are related to the spatial ones by $\partial_0 \gamma^{00} = -\partial_i \gamma^{i0}$. We are thus left with

$$
\widetilde{\gamma}_{ij}(\omega, \boldsymbol{x}) = 4G \frac{e^{-i\omega r}}{r} \int d^3 y\, \widetilde{T}^{ij}(\omega, \boldsymbol{y}).
\tag{S84}
$$

(c) We now use the identity $i\omega \widetilde{T}^{0\nu} = -\partial_i \widetilde{T}^{i\nu}$ to relate the spatial components of the energy-momentum tensor to the time components

$$
\begin{aligned}
\int d^3 y\, \widetilde{T}^{ij}(\omega, \boldsymbol{y}) &= \int d^3 y \left(\partial_k (\widetilde{T}^{ik} y^j) - y^j \partial_k \widetilde{T}^{ik} \right) \\
&= - \int d^3 y\, y^j \partial_k \widetilde{T}^{ik} \\
&= i\omega \int d^3 y\, y^j \widetilde{T}^{i0}.
\end{aligned}
\tag{S85}
$$

The first step is just rewriting with the product rule (note that $\partial_k y^j = \delta_k^j$), the second step is a consequence of the divergence theorem, and in the third step we used the conservation of the energy-momentum tensor.

We know that the metric perturbation $\widetilde{\gamma}_{ij}$ is symmetric, so the right-hand side expression in the above equation must be as well. We thus rewrite

$$
i\omega \int d^3 y\, y^j \widetilde{T}^{i0} = \frac{i\omega}{2} \int d^3 y \left(y^j \widetilde{T}^{i0} + y^i \widetilde{T}^{j0} \right).
\tag{S86}
$$

Now we can use the same trick again and rewrite

$$\frac{i\omega}{2} \int d^3y \left(y^i \widetilde{T}^{j0} + y^j \widetilde{T}^{i0} \right)$$

$$= \frac{i\omega}{2} \int d^3y \left(\partial_s (y^i y^j \widetilde{T}^{s0}) - y^i y^j \partial_s \widetilde{T}^{s0} - \widetilde{T}^{j0} y^i + \widetilde{T}^{j0} y^i \right)$$

$$= \frac{i\omega}{2} \int d^3y \left(-y^i y^j \partial_s \widetilde{T}^{s0} \right) \tag{S87}$$

$$= -\frac{\omega^2}{2} \int d^3y \, y^i y^j \widetilde{T}^{00} \, ,$$

where in the first step we again rewrote $y^i \widetilde{T}^{j0}$ using the product rule, the second step is divergence theorem and the third is using the conservation law. We now define the quadrupole moment tensor

$$I_{ij}(t) = \int d^3y \, y^i y^j T^{00}(t, \mathbf{y}) \, . \tag{S88}$$

We are thus finally left with

$$\widetilde{\gamma}_{ij}(\omega, \mathbf{x}) = -4G \frac{e^{-i\omega r}}{r} \frac{\omega^2}{2} \widetilde{I}_{ij}(\omega). \tag{S89}$$

(d) Let us do the inverse Fourier transform

$$\gamma_{ij}(t, \mathbf{x}) = -\frac{2G}{r} \frac{1}{\sqrt{2\pi}} \int d\omega e^{i\omega t} e^{-i\omega r} \omega^2 \widetilde{I}_{ij}(\omega)$$

$$= -\frac{2G}{r} \frac{1}{\sqrt{2\pi}} \int d\omega e^{i\omega t_r} \omega^2 \widetilde{I}_{ij}(\omega)$$

$$= \frac{2G}{r} \frac{1}{\sqrt{2\pi}} \int d\omega \frac{d^2}{dt^2} \left(e^{i\omega t_r} \right) \widetilde{I}_{ij}(\omega) \tag{S90}$$

$$= \frac{2G}{r} \frac{d^2}{dt^2} \left(\frac{1}{\sqrt{2\pi}} \int d\omega e^{i\omega t_r} \widetilde{I}_{ij}(\omega) \right)$$

$$= \frac{2G}{r} \frac{d^2 I_{ij}}{dt^2} \Big|_{t_r} \, .$$

We find the quadrupole radiation formula

$$\gamma_{ij}(t, \mathbf{x}) = \frac{2G}{r} \frac{d^2 I_{ij}}{dt^2} \Big|_{t_r} \, . \tag{S91}$$

Note that to lowest order the gravitational radiation is given by a changing quadrupole moment of the source. In contrast, the leading order electromagnetic radiation arises from a changing dipole moment of the charge density. The difference comes from the nature of dipole radiation: It corresponds to a motion of the center of density—charge density in the electromagnetic case. However, oscillation of the center of mass of an isolated system violates conservation of momentum, so the leading order term of gravitational radiation arises from a change in the quadrupole moment.

Problem 3.4 *(Gravitational Bremsstrahlung)*

The parabolic orbit of the mass m scattering on $M \gg m$ is described by the parametric form

$$\vec{r}(\varphi) = \begin{pmatrix} x \\ y \\ z \end{pmatrix} = \frac{2b}{1 + \cos(\varphi)} \begin{pmatrix} \cos(\varphi) \\ \sin(\varphi) \\ 0 \end{pmatrix}, \qquad (S92)$$

where the time dependence is described by

$$\dot{\varphi} = \sqrt{\frac{GM}{8b^3}} \, [1 + \cos(\varphi)]^2 \, . \qquad (S93)$$

Thus we have for the quadrupole tensor

$$I_{ij}(t) = \int d^3 y' \rho(\vec{y}') y_i' y_j' = \frac{4b^2 m}{\left(1 + \cos(\varphi)\right)^2} \begin{pmatrix} \cos^2(\varphi) & \sin(\varphi)\cos(\varphi) & 0 \\ \sin(\varphi)\cos(\varphi) & \sin^2(\varphi) & 0 \\ 0 & 0 & 0 \end{pmatrix}, \qquad (S94)$$

where we have used $\rho(\vec{y}') = M\delta(\vec{y}') + m\delta(\vec{y}' - \vec{r}(\varphi))$. Note that the angular position φ is not integrated over since it is given by the Newtonian equation of motion for m. Now we can use that for any function $f(\varphi)$ and $\dot{\varphi} = \dot{\varphi}(\varphi)$ we have

$$\frac{d^2 f(\varphi)}{dt^2} = \frac{d^2\varphi}{dt^2}\frac{df}{d\varphi} + \frac{d^2 f}{d\varphi^2}\left(\frac{d\varphi}{dt}\right)^2 \qquad (S95)$$

$$= \dot{\varphi}\frac{d\dot{\varphi}}{d\varphi}\frac{df}{d\varphi} + \dot{\varphi}^2\frac{d^2 f}{d\varphi^2} \qquad (S96)$$

$$= \frac{1}{2}\frac{d\dot{\varphi}^2}{d\varphi}\frac{df}{d\varphi} + \dot{\varphi}^2\frac{d^2 f}{d\varphi^2} \, . \qquad (S97)$$

With the above relation and Eq. (S93) the second derivative of the quadrupole tensor simplifies to

$$\frac{d^2 I_{ij}}{dt^2} = \frac{GM}{8b^3}\left(-2\sin(\varphi)\left(1 + \cos(\varphi)\right)^3\frac{dI_{ij}}{d\varphi} + (1 + \cos(\varphi))^4\frac{d^2 I_{ij}}{d\varphi^2}\right). \qquad (S98)$$

For the first and second term in the above equation we obtain

$$\sin(\varphi)\big(1+\cos(\varphi)\big)^3 \frac{dI_{ij}}{dt} =$$

$$4b^2 \begin{pmatrix} -2\sin^2(\varphi)\cos(\varphi) & \sin(\varphi)\left(\cos(2\varphi)+\cos(\varphi)\right) & 0 \\ \sin(\varphi)\left(\cos(2\varphi)+\cos(\varphi)\right) & 2\sin^2(\varphi)(\cos(\varphi)+1) & 0 \\ 0 & 0 & 0 \end{pmatrix}, \tag{S99}$$

and

$$\big(1+\cos(\varphi)\big)^4 \frac{d^2 I_{ij}}{dt^2} =$$

$$8b^2 \begin{pmatrix} \cos^2\left(\frac{\varphi}{2}\right)(-6\cos(\varphi)+\cos(2\varphi)+3) & 2\left(\sin\left(\frac{3\varphi}{2}\right)-5\sin\left(\frac{\varphi}{2}\right)\right)\cos^3\left(\frac{\varphi}{2}\right) & 0 \\ 2\left(\sin\left(\frac{3\varphi}{2}\right)-5\sin\left(\frac{\varphi}{2}\right)\right)\cos^3\left(\frac{\varphi}{2}\right) & -4\cos^4\left(\frac{\varphi}{2}\right)(\cos(\varphi)-2) & 0 \\ 0 & 0 & 0 \end{pmatrix}. \tag{S100}$$

Finally we have

$$\gamma_{ij} = \frac{2G^2 mM}{rbc^4} \begin{pmatrix} \left(\sin^2(\varphi)-\cos^2(\varphi)(\cos(\varphi)+1)\right) & -8\sin\left(\frac{\varphi}{2}\right)\cos^5\left(\frac{\varphi}{2}\right) & 0 \\ -8\sin\left(\frac{\varphi}{2}\right)\cos^5\left(\frac{\varphi}{2}\right) & 4\cos^4\left(\frac{\varphi}{2}\right)\cos(\varphi) & 0 \\ 0 & 0 & 0 \end{pmatrix}. \tag{S101}$$

We immediately see that the metric perturbation in the trace-reversed Lorenz gauge is not traceless. The Lorenz gauge condition relates the time components to the spatial components

$$k^\mu \gamma_{\mu\nu} = k(\gamma_{0\nu} - \gamma_{3\nu}) = 0. \tag{S102}$$

Thus we have $\gamma_{0i} = \gamma_{3i} = 0$ and $\gamma_{00} = \gamma_{03} = 0$.

In the vacuum we can project to transverse traceless gauge

$$h_{ij}^{TT} = \Lambda_{ij}^{kl} \gamma_{kl}, \tag{S103}$$

where the projection tensor is given by

$$\Lambda_{ij}^{kl} = P_i^k P_j^l - \frac{1}{2} P_{ij} P^{kl}. \tag{S104}$$

Using $P_i^j = \delta_i^j - n_i n^j$ we have for an observer on the z-axis with $\vec{n} = (0, 0, 1)$

$$h_{11}^{TT} = \frac{1}{2}(\gamma_{11} - \gamma_{22}), \quad h_{22}^{TT} = \frac{1}{2}(\gamma_{22} - \gamma_{11}) = -h_{11}^{TT}, \quad h_{12}^{TT} = h_{21}^{TT} = \gamma_{12}, \tag{S105}$$

and all the other components vanish. Figure 5.2 shows the time dependence of the two independent polarizations $h_+ = h_{11}^{TT}$ and $h_\times = h_{12}^{TT}$.

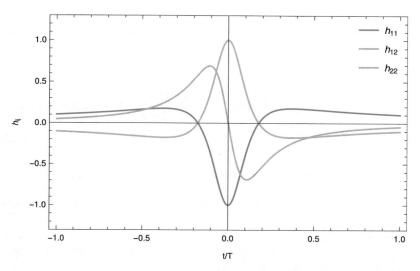

Fig. S.2 Time dependence of the h_{11}^{TT} and h_{22}^{TT} h_{12}^{TT} components normalized to $h_{11}(\varphi = 0)$. The time is expressed in terms of T_{o}

The timescale of the transit is given by the orbital time $T_{\mathrm{o}} = 2\pi\sqrt{\frac{b^3}{GM}}$, whereas the amplitude is affected by $T_{\mathrm{gw}}^2/T_{\mathrm{o}}^2$, where we defined

$$T_{\mathrm{gw}} = \sqrt{\frac{Gmb^2}{rc^4}}\,, \tag{S106}$$

and the time dependence of the orbit was obtained solving the implicit equation for $\varphi(t)$

$$t(\varphi) = \sqrt{\frac{2b^3}{GM}}\left(\tan\left(\frac{\varphi}{2}\right) + \frac{1}{3}\tan^3\left(\frac{\varphi}{2}\right)\right). \tag{S107}$$

Problem 3.5 *(A spinning rod)*

(a) The time-dependent parts of the reduced quadrupole moment tensor are

$$Q_{xx} = \mu\cos^2(\omega t)\cdot 2\int_0^{l/2} x^2 \mathrm{d}x = \frac{\mu l^3}{12}\cos^2\omega t = \frac{Ml^2}{24}\cos(2\omega t) + C_1\,, \tag{S108}$$

$$Q_{yy} = -\frac{Ml^2}{24}\cos 2\omega t + C_2\,, \tag{S109}$$

$$Q_{xy} = Q_{yx} = \frac{Ml^2}{24} \sin 2\omega t. \tag{S110}$$

(b)

$$P_{GW} = L_{GW} = \frac{G}{5c^5} \langle \dddot{Q}_{jk} \dddot{Q}_{jk} \rangle = \frac{(2\omega)^6 \, G}{5c^5} \left(\frac{Ml^2}{24} \right)^2 \langle 2\cos^2 2\omega t + 2\sin^2 2\omega t \rangle$$

$$= \frac{2G}{45c^5} \omega^6 M^2 l^4, \tag{S111}$$

(c) Balancing the electrostatic and centripetal forces on the charges,

$$e \mid \nabla\phi \mid = rm\omega^2, \tag{S112}$$

where ϕ is the electrostatic potential, and e and m are the charge and mass of an electron, respectively. Maxwell's Gauss's law equation tells us that the negative gradient of the potential is the density, so

$$\frac{\rho}{\varepsilon_0} \sim -\nabla^2 \phi \sim -\frac{m}{e}\omega^2. \tag{S113}$$

So over the whole volume lA, the charge is $(m/e)\,\varepsilon_0\omega^2 lA$, where A is the rod's x-section. Thus the rod becomes an electric quadrupole with quadrupole moment $Q_E \sim (m/e)\,\varepsilon_0\omega^2 l^3 A$.

(d) The electric quadrupole radiation due to this induced moment is approximately $\omega^6 Q_E^2/(\varepsilon_0 c^5)$, thus

$$L_{EM} \sim \frac{\varepsilon_0}{c^5}\omega^{10} \left(\frac{m}{e} \right)^2 l^6 A^2. \tag{S114}$$

(e) The ratio of electromagnetic damping to gravitational damping is then

$$\frac{L_{EM}}{L_{GW}} \sim \frac{45}{2} \frac{\varepsilon_0 c^5 \omega^{10} m^2 l^6 A^2}{Gc^5 e^2 \omega^6 M^2 l^4} = \frac{45}{2} \frac{\varepsilon_0 m^2 \omega^4 l^2 A^2}{Ge^2 M^2} = \frac{45}{2} \frac{\varepsilon_0 m^2 \omega^4}{Ge^2 \rho^2}. \tag{S115}$$

Plugging in the numbers we find:

$$\frac{L_{EM}}{L_{GW}} \sim 10^{-17}. \tag{S116}$$

Problem 3.6 *(Order of magnitude estimates)*

(a) We know $T(t = 0)$ and $m_1 + m_2 = M$, the time to coalescence t_{coal} is given according to Eq. (3.124), and we find:

$$t_{\text{coal}} = \frac{5}{256} \frac{c^5 M^{1/3}}{G^{5/3} m_1 m_2} \left(\frac{T}{2\pi}\right)^{8/3} \approx 5.4 \times 10^{14} \text{s} \approx 1.7 \times 10^7 \text{year}. \quad \text{(S117)}$$

(b) As the two black holes are merging, we are in the limiting case where the Newtonian approximation breaks down. This is typically the case when the two black holes are separated by a distance comparable to their Schwarzschild radius, i.e., $R \approx 2GM/c^2$.

(i) The frequency of the gravitational waves (twice the orbital frequency) is approximately:

$$\nu = \frac{2\omega}{2\pi} = \frac{1}{\pi}\sqrt{\frac{GM}{R^3}} = \frac{1}{\pi\sqrt{8}}\frac{c^3}{GM} \approx 10^{-2} \text{ Hz}. \quad \text{(S118)}$$

(ii) The dimensionless strain, defined as the relative amplitude of the oscillation of a ruler when a gravitational wave passes through, is:

$$h = |h_{ij}| = \frac{GM}{c^2 |x|} \approx 10^{-16}. \quad \text{(S119)}$$

(iii) We take here the formula from part (a), with the frequency of part (i):

$$t = \frac{5}{256} \frac{2^{1/3} c^5}{(GM)^{5/3}} \left(\frac{1}{\pi\nu}\right)^{8/3} \approx 20 \text{ s}. \quad \text{(S120)}$$

Problem 3.7 *(Gravitational waves from a binary system)*

(a) Let us solve the problem of circular orbits in Newtonian gravity. We have that the two bodies m_1 and m_2 are following a circular orbit around a common center, with radius, respectively, r_1 and r_2, in the x-y plane. Their positions are:

$$x_1^i = r_1(\cos\omega t, \sin\omega t, 0), \quad x_2^i = -r_2(\cos\omega t, \sin\omega t, 0).$$

Newton's second law gives:

$$\frac{Gm_1 m_2}{(r_1 + r_2)^2} = m_1 r_1 \omega^2 = m_2 r_2 \omega^2.$$

We find, with $R = r_1 + r_2$ and $M = m_1 + m_2$:

$$\frac{r_2}{m_1} = \frac{r_1}{m_2} = \frac{R}{M}$$

$$\omega^2 = \frac{GM}{R^3}.$$

The density is:

$$\rho\left(t, x^i\right) = m_1\delta\left(x^i - x_1^i(t)\right) + m_2\delta\left(x^i - x_2^i(t)\right) .$$

So that the quadrupole moment is:

$$I_{ij}(t) = \int \rho\left(t, y\right) y^i y^j d^3y = m_1 x_1^i(t) x_1^j(t) + m_2 x_2^i(t) x_2^j(t)$$

$$I_{xx} = m_1 r_1^2 \cos^2 \omega t + m_2 r_2^2 \cos^2 \omega t = \mu R^2 \cos^2 \omega t$$

$$I_{xy} = m_1 r_1^2 \cos \omega t \sin \omega t + m_2 r_2^2 \cos \omega t \sin \omega t = \mu R^2 \cos \omega t \sin \omega t$$

$$I_{yy} = m_1 r_1^2 \sin^2 \omega t + m_2 r_2^2 \sin^2 \omega t = \mu R^2 \sin^2 \omega t .$$

where $\mu = \frac{m_1 m_2}{M}$.

The gravitational wave tensor is thus far away from the source and on the z-axis:

$$\gamma_{ij}(t, x^i) = \frac{2G}{|x|}\frac{d^2}{dt^2} I_{ij}(t - |x|)$$

$$\gamma_{xx}(t, x^i) = \frac{4G\mu R^2\omega^2}{z}\left(\sin^2 \omega t_r - \cos^2 \omega t_r\right) = -\frac{4G^2\mu M}{Rz}\cos 2\omega t_r = -\gamma_{yy}(t, x^i)$$

$$\gamma_{xy}(t, x^i) = -\frac{8G\mu R^2\omega^2}{z}\sin \omega t_r \cos \omega t_r = -\frac{4G^2\mu M}{Rz}\sin 2\omega t_r ,$$

where $t_r = t - z$ is the retarded time.
(b) The energy loss is given by (see Eq. (3.107), with $J_{ij} = Q_{ij}$ and $c = 1$)

$$P = -\frac{G}{5}\left\langle \frac{d^3 J_{ij}}{dt^3} \frac{d^3 J^{ij}}{dt^3} \right\rangle, \tag{S121}$$

where

$$J_{ij} = I_{ij} - \frac{1}{3}\delta_{ij}\delta^{kl}I_{kl} \quad \text{and} \quad J^{ij} = \eta^{i\alpha}\eta^{j\beta}J_{\alpha\beta}. \tag{S122}$$

The nonzero components of the third derivative of the reduced quadrupole moment tensor are

$$\frac{d^3 J^{xx}}{dt^3} = 4\mu R^2\omega^3 \sin(2\omega t) = -\frac{d^3 J^{yy}}{dt^3}, \quad \frac{d^3 J^{xy}}{dt^3} = -4\mu R^2\omega^3 \cos(2\omega t) = \frac{d^3 J^{yx}}{dt^3}. \tag{S123}$$

Substituting into (S121) and averaging the trig functions over one oscillation gives

$$P = -\frac{32}{5}G\mu^2 R^4\omega^6 \quad \text{or} \quad P = -\frac{32}{5}G^4\frac{\mu^2 M^3}{R^5}. \tag{S124}$$

(c) The Newtonian energy of the binary is the following:

$$E_N = E_{kin} + E_{pot} = \frac{1}{2}\left(m_1 r_1^2 \omega^2 + m_2 r_2^2 \omega^2\right) - \frac{Gm_1 m_2}{R} = -\frac{1}{2}\frac{G\mu M}{R}.$$

We thus have:

$$\frac{dE_N}{dt} = \frac{1}{2}\frac{G\mu M}{R^2}\frac{dR}{dt} = -\frac{dE_{GW}}{dt}.$$

From this, we can infer:

$$\frac{dR}{dt} = -\frac{64G^3\mu M^2}{5R^3}.$$

(d) We have ω in terms of R:

$$\omega = \sqrt{\frac{GM}{R^3}} \quad \Longrightarrow \quad \frac{d\omega}{dt} = -\frac{3}{2}\sqrt{\frac{GM}{R^5}}\frac{dR}{dt} = \frac{96}{5}\sqrt{\frac{G^7\mu^2 M^5}{R^{11}}} = \frac{96}{5}G^{5/3}\mu M^{2/3}\omega^{11/3}.$$

We thus find that the derivative of the frequency depends on $\mathcal{M}^{5/3}$, where $\mathcal{M}$ is the chirp mas defined as: $\mathcal{M} = \mu^{3/5}M^{2/5}$.

(e) We can integrate $\omega(t)$:

$$\omega^{-11/3}\frac{d\omega}{dt} = \frac{96}{5}(G\mathcal{M})^{5/3} \quad \Longrightarrow \quad \omega(t) = \left(\frac{5}{256}\right)^{3/8}(G\mathcal{M})^{-5/8}(t_0 - t)^{-3/8},$$

where t_0 is a constant.

This diverges as $t \to t_0$, so the coalescence time is $T_{coal} = t_0$, with

$$t_0 = \frac{5}{2^8}\frac{1}{(G\mathcal{M})^{5/3}\omega^{8/3}}. \tag{S125}$$

Problem 3.8 (*Post-Newtonian–Lagrangian for a binary system*)

(a) The action is given by

$$S_1 = -m_1 \int dt \left(-g_{00}c^2 - 2cg_{0i}v_1^i - g_{ij}v_1^i v_1^j\right)^{1/2} =$$

$$= -m_1 c^2 \int dt \left(1 - {}^{(2)}g_{00} - {}^{(4)}g_{00} - 2{}^{(3)}g_{0i}\frac{v_1^i}{c} - \frac{v_1^2}{c^2} - {}^{(2)}g_{ij}\frac{v_1^i v_1^j}{c^2}\right)^{1/2}. \tag{S126}$$

We can expand the big parenthesis in equation (S126) as $(1+x)^{1/2} \simeq 1 + x/2 - x^2/8 + \ldots$, and keeping only terms up to ϵ^4 we have:

$$S_1 = -m_1 c^2 \int dt \left\{ \left(1 - \frac{{}^{(2)}g_{00}}{2} - \frac{v_1^2}{2c^2} \right) \right.$$

$$\left. + \left(-\frac{v_1^4}{8c^4} - \frac{{}^{(2)}g_{00}^2}{8} - \frac{{}^{(2)}g_{00}}{4} \frac{v_1^2}{c^2} - \frac{{}^{(4)}g_{00}}{2} - \frac{{}^{(3)}g_{0i} v_1^i}{c} - \frac{{}^{(2)}g_{ij} v_1^i v_1^j}{2c^2} \right) \right\}$$

$$= -m_1 c^2 \int dt + m_1 c^2 \int dt \left\{ \left(\frac{{}^{(2)}g_{00}}{2} + \frac{v_1^2}{2c^2} \right) \right.$$

$$\left. + \left(+\frac{v_1^4}{8c^4} + \frac{{}^{(2)}g_{00}^2}{8} + \frac{{}^{(2)}g_{00}}{4} \frac{v_1^2}{c^2} + \frac{{}^{(4)}g_{00}}{2} + \frac{{}^{(3)}g_{0i} v_1^i}{c} + \frac{{}^{(2)}g_{ij} v_1^i v_1^j}{2c^2} \right) \right\}.$$

$$\text{(S127)}$$

The first integral in equation (S127) is associated with the rest mass energy and does not appear in any equation. We can neglect it. The second integral contains the actual Lagrangian of the particle m_1. The first term inside the second integral is the Newtonian part $(\mathcal{O}(\epsilon^2))$, whereas the second term is the post-Newtonian part $(\mathcal{O}(\epsilon^4))$. Using the expressions for ${}^{(i)}g_{\mu\nu}$, in terms of the potentials ϕ, ξ and ψ we find:

$$\mathcal{L} = m_1 c^2 \left\{ \frac{1}{2} \left(\frac{v_1}{c} \right)^2 - \phi + \frac{1}{8} \left(\frac{v_1}{c} \right)^4 - \frac{\phi^2}{2} - \psi - \frac{3\phi}{2} \left(\frac{v_1}{c} \right)^2 + \xi_i \frac{v_1^i}{c} \right\}.$$

$$\text{(S128)}$$

(b) The motion of particle m_1 is derived from the contribution of m_2 only to the metric perturbations ${}^{(i)}g_{\mu\nu}$. This means in turn that $T^{\mu\nu}$ from which ${}^{(i)}g_{\mu\nu}$ are derived depends only on m_2:

$$T^{\mu\nu} = \frac{1}{\sqrt{-g}} \frac{dt}{d\tau_2} m_2 \frac{dx_2^\mu}{dt} \frac{dx_2^\nu}{dt} \delta^{(3)} (\mathbf{x} - \mathbf{x}_2(t)) . \qquad \text{(S129)}$$

From $-g \simeq 1 - 4\phi$, it follows $(-g)^{-1/2} \simeq 1 + 2\phi$, and thus:

$$\frac{d\tau_2}{dt} = \frac{1}{c} \left(-g_{\mu\nu} \frac{dx_2^\mu}{dt} \frac{dx_2^\nu}{dt} \right)^{1/2} \simeq 1 + \phi - \frac{v_2^2}{2c^2}, \quad \Rightarrow \quad \frac{dt}{d\tau_2} \simeq 1 - \phi + \frac{v_2^2}{2c^2}.$$

$$\text{(S130)}$$

Putting everything together we get:

$$T^{\mu\nu} = \left(1 + \phi + \frac{v_2^2}{2c^2} \right) m_2 \frac{dx_2^\mu}{dt} \frac{dx_2^\nu}{dt} \delta^{(3)} (\mathbf{x} - \mathbf{x}_2(t)) , \qquad \text{(S131)}$$

which can be finally expanded as:

$$^{(0)}T^{00} = m_2 c^2 \delta^{(3)} \left(\mathbf{x} - \mathbf{x}_2(t) \right), \tag{S132}$$

$$^{(2)}T^{00} = m_2 c^2 \left(\phi + \frac{v_2^2}{2c^2} \right) \delta^{(3)} \left(\mathbf{x} - \mathbf{x}_2(t) \right), \tag{S133}$$

$$^{(1)}T^{0i} = m_2 c^2 \frac{v_2^i}{c} \delta^{(3)} \left(\mathbf{x} - \mathbf{x}_2(t) \right), \tag{S134}$$

$$^{(2)}T^{ij} = m_2 c^2 \frac{v_2^i v_2^j}{c^2} \delta^{(3)} \left(\mathbf{x} - \mathbf{x}_2(t) \right). \tag{S135}$$

(c) Let us start with ϕ. We have:

$$\phi(\mathbf{x}, t) = -\frac{G}{c^2} \int \frac{d\mathbf{x}' m_2 \delta^{(3)} \left(\mathbf{x}' - \mathbf{x}_2(t) \right)}{|\mathbf{x}' - \mathbf{x}_2(t)|} = -\frac{Gm_2}{|\mathbf{x} - \mathbf{x}_2(t)| \, c^2}. \tag{S136}$$

For ξ_i we have:

$$\xi_i(\mathbf{x}, t) = -\frac{4G}{c^2} \frac{v_2^i}{c} \int \frac{d\mathbf{x}' m_2 \delta^{(3)} \left(\mathbf{x}' - \mathbf{x}_2(t) \right)}{|\mathbf{x}' - \mathbf{x}_2(t)|} = -\frac{4Gm_2}{|\mathbf{x} - \mathbf{x}_2(t)| \, c^2} \frac{v_2^i}{c}, \tag{S137}$$

or:

$$\xi(\mathbf{x}, t) = -\frac{4Gm_2}{|\mathbf{x} - \mathbf{x}_2(t)| \, c^2} \frac{\mathbf{v}_2}{c}. \tag{S138}$$

Finally, we have to evaluate ψ, which is:

$$\psi(\mathbf{x}, t) = -\frac{G}{c^2} \int \frac{d\mathbf{x}' m_2 \delta^{(3)} \left(\mathbf{x}' - \mathbf{x}_2(t) \right)}{|\mathbf{x}' - \mathbf{x}_2(t)|} \left(\frac{3}{2} \frac{v_2^2}{c^2} + \phi \right) - \int \frac{d\mathbf{x}'}{4\pi \, |\mathbf{x}' - \mathbf{x}(t)|} \partial_0^2 \phi. \tag{S139}$$

Calculating the integrals yields:

$$\psi(\mathbf{x}, t) = \frac{Gm_2}{2 \, |\mathbf{x} - \mathbf{x}_2(t)| \, c^2} \tag{S140}$$

$$\cdot \left\{ -3 \frac{v_2^2}{c^2} + \frac{\mathbf{v}_1 \cdot \mathbf{v}_2}{c} - \frac{[(\mathbf{x} - \mathbf{x}_2(t)) \cdot \mathbf{v}_1] \, [(\mathbf{x} - \mathbf{x}_2(t)) \cdot \mathbf{v}_2]}{c \, |\mathbf{x} - \mathbf{x}_2(t)|^2} \right\}. \tag{S141}$$

(d) Putting all the expressions for the potentials into equation (S128) and applying the generic position $\mathbf{x}$ to $\mathbf{x}_1$ we get the Lagrangian for m_1 up to 1 PN correction:

$$\mathcal{L} = \frac{1}{2}m_1 v_1^2 + \frac{Gm_1 m_2}{r_{12}} + \frac{1}{8}m_1 \frac{v_1^4}{c^2} - \frac{G^2 m_2^2 m_1}{2r_{12}^2 c^2}$$
$$+ \frac{Gm_1 m_2}{2r_{12}} \left\{ 3\left[\frac{v_1^2}{c^2} + \frac{v_2^2}{c^2}\right] - 7\frac{\mathbf{v}_1 \cdot \mathbf{v}_2}{c^2} - \frac{(\hat{\mathbf{r}}_{12} \cdot \mathbf{v}_1)(\hat{\mathbf{r}}_{12} \cdot \mathbf{v}_2)}{c^2} \right\}, \quad \text{(S142)}$$

where we defined $r_{12} = |\mathbf{x}_1 - \mathbf{x}_2|$ and $\hat{\mathbf{r}}_{12} = (\mathbf{x}_1 - \mathbf{x}_2) / |\mathbf{x}_1 - \mathbf{x}_2|$. There are two purely symmetric terms in the subscripts 1 and 2, which are $Gm_1 m_2 / r_{12}$ and $Gm_1 m_2 / (2r_{12}) \{\ldots\}$. Therefore, summing up the above Lagrangian with the equivalent one for m_2 without double counting the symmetrical terms gives finally:

$$\mathcal{L} = \frac{1}{2}m_1 v_1^2 + \frac{1}{2}m_2 v_2^2 + \frac{Gm_1 m_2}{r_{12}} + \frac{1}{8}m_1 \frac{v_1^4}{c^2} + \frac{1}{8}m_2 \frac{v_2^4}{c^2} - \frac{G^2 m_2 m_1 (m_1 + m_2)}{2r_{12}^2 c^2}$$
$$+ \frac{Gm_1 m_2}{2r_{12}} \left\{ 3\left[\frac{v_1^2}{c^2} + \frac{v_2^2}{c^2}\right] - 7\frac{\mathbf{v}_1 \cdot \mathbf{v}_2}{c^2} - \frac{(\hat{\mathbf{r}}_{12} \cdot \mathbf{v}_1)(\hat{\mathbf{r}}_{12} \cdot \mathbf{v}_2)}{c^2} \right\}. \quad \text{(S143)}$$

Problem 3.9 (*Hamiltonian geodesic formulation in PN framework*)

(a) The Lagrangian for the full theory is

$$L = \frac{1}{2}\left[-\left(1 - \frac{r_s}{r}\right)\dot{t}^2 + \left(1 - \frac{r_s}{r}\right)^{-1}\dot{r}^2 + r^2\dot{\theta}^2 + r^2 \sin^2\theta \,\dot{\phi}^2 \right]. \quad \text{(S144)}$$

The Hamiltonian for the full theory is

$$\mathcal{H} = \frac{1}{2}\left[-p_t^2\left(1 - \frac{r_s}{r}\right)^{-1} + p_r^2\left(1 - \frac{r_s}{r}\right) + \frac{p_\theta^2}{r^2} + \frac{p_\phi^2}{r^2 \sin^2\theta} \right]. \quad \text{(S145)}$$

(b) This gives us the equation of motion:

$$\dot{t} = -p_t\left(1 - \frac{r_s}{r}\right)^{-1}, \quad \text{(S146)}$$

$$\dot{r} = p_r\left(1 - \frac{r_s}{r}\right), \quad \text{(S147)}$$

$$\dot{\theta} = \frac{p_\theta}{r^2}, \quad \text{(S148)}$$

$$\dot{\phi} = \frac{p_\phi}{r^2 \sin^2\theta}, \quad \text{(S149)}$$

$$\dot{p}_t = 0, \quad \text{(S150)}$$

$$\dot{p}_r = \frac{r_s}{r^2}\left(-\frac{p_t^2}{2}\left(1 - \frac{r_s}{r}\right)^{-2} - \frac{p_r^2}{2} + \frac{p_\theta^2}{rr_s} + \frac{p_\phi^2}{rr_s \sin^2\theta} \right), \quad \text{(S151)}$$

$$\dot{p}_\theta = \frac{p_\phi^2 \cot\theta}{r^2 \sin^2\theta}, \tag{S152}$$

$$\dot{p}_\phi = 0. \tag{S153}$$

(c) Let us introduce a parameter ϵ to help us keep track of our orders. Let us choose $r_s = \mathcal{O}(\epsilon^2)$. For a massive particle we have $v \ll c$, so we find $p_t = \mathcal{O}(1)$, $p_r, p_\theta, p_\phi = \mathcal{O}(\epsilon)$, while for a massless particle we have $p_\mu = \mathcal{O}(1)$. To expand the equations of motion we, e.g., make the replacements $r_s \to \epsilon^2 r_s, p_r \to \epsilon p_r, p_\theta \to \epsilon p_\theta, p_\phi \to \epsilon p_\phi$ and then expand to second order around $\epsilon = 0$. Applying this procedure we find the following equations of motion:

	massless	massive	
$\dot{t}$	$\dot{t} = -p_t \left(1 + \dfrac{r_s}{r}\right)$	$\dot{t} = -p_t \left(1 + \dfrac{r_s}{r}\right)$	(S154)
	$\dot{r} = p_r \left(1 - \dfrac{r_s}{r}\right)$	$\dot{r} = p_r$	(S155)
	$\dot{\theta} = \dfrac{p_\theta}{r^2}$	$\dot{\theta} = \dfrac{p_\theta}{r^2}$	(S156)
	$\dot{\phi} = \dfrac{p_\phi}{r^2 \sin^2\theta}$	$\dot{\phi} = \dfrac{p_\phi}{r^2 \sin^2\theta}$	(S157)
	$\dot{p}_t = 0$	$\dot{p}_t = 0$	(S158)

massless:
$$\dot{p}_r = -\frac{p_t^2 r_s}{2r^2} - \frac{p_r^2 r_s}{2r^2} + \frac{p_\theta^2}{r^3} + \frac{p_\phi^2}{r^3 \sin^2\theta}$$

massive:
$$\dot{p}_r = -\frac{p_t^2 r_s}{2r^2} + \frac{p_\theta^2}{r^3} + \frac{p_\phi^2}{r^3 \sin^2\theta} \tag{S159}$$

massless:
$$\dot{p}_\theta = \frac{p_\phi^2 \cot\theta}{r^2 \sin^2\theta}$$

massive:
$$\dot{p}_\theta = \frac{p_\phi^2 \cot\theta}{r^2 \sin^2\theta} \tag{S160}$$

massless:
$$\dot{p}_\phi = 0$$

massive:
$$\dot{p}_\phi = 0. \tag{S161}$$

(d) To plot trajectories we first fix the orbital plane by setting $\theta = \pi/2$ and therefore $p_\theta = 0$. Then we can fix two more constants, p_t corresponding to energy and p_ϕ corresponding to angular momentum. Afterward we can integrate the equations for $\dot{r}$, $\dot{\phi}$ and $\dot{p}_r$ numerically with suitable initial conditions. Knowing r and ϕ we can plot some trajectories in the orbital plane. We can see that the closer to the horizon we come the more the approximate solution deviates from the full solution. Note also that the full solution breaks down at the horizon due to the coordinate singularity of Schwarzschild coordinates, whereas the approximate solution is just physically invalid near the horizon.

Fig. S.3 Trajectory of two
photons, each with the same
impact parameter around a
Schwarzschild black hole.
Both photons have the same
initial conditions, yet the red
photon has been calculated
using the full metric, and the
blue the PN metric. Black
shows the event horizon
surface. Note that the red
photon grazes the *photon
sphere* at $r = 3r_s$

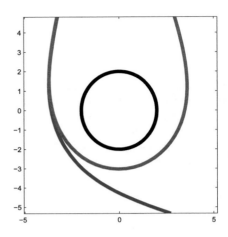

Solution to Problem in Chap. 4

Problem 4.1 *(In-falling object into a black hole)*

(a) The object falls freely and follows a radial geodesic, i.e., a geodesic satisfying
$u^\theta = u^\phi = 0$. Indeed, a geodesic starting from $x_*^\nu = (t_*, r_*, \theta_*, \phi_*)$ with vanish-
ing initial velocity $\dot{x}^i(x_*^\nu) = 0$ will not have any reason to prefer any direction
for θ and ϕ, due to the spherical symmetry.

The Schwarzschild metric can be written as:

$$ds^2 = -B(r)dt^2 + \frac{1}{B(r)}dr^2 + r^2 d\omega^2, \tag{S162}$$

$$B(r) = 1 - \frac{2GM}{r}, \qquad d\omega^2 = d\theta^2 + \sin^2\theta d\phi^2. \tag{S163}$$

From here on we use the notation $B(r) = B$ and $B(r_*) = B_*$. To get the radial
geodesic, we do not need the full geodesic equation. We start with

$$-1 = -B(u^t)^2 + \frac{1}{B}(u^r)^2, \tag{S164}$$

for a radial geodesic. We also know the integral of motion $Bu^t = F$ and the initial
condition $u^r(r_*) = 0$. Evaluation of this equation at $r = r_*$ gives us the condition
$F^2 = B_*$. Inserting this back into the equation we get

$$-1 = -\frac{B_*}{B} + \frac{1}{B}(u^r)^2. \tag{S165}$$

We solve for u^r and get $(u^r)^2 = (B_* - B)$. With this we can also solve for u^t and get $(u^t)^2 = B_*/B^2$. The inward radial geodesic is thus given by the 4-velocity

$$u^\mu = \left(\frac{\sqrt{B_*}}{B}, -\sqrt{B_* - B}, 0, 0 \right). \tag{S166}$$

Thus, we can compute:

$$\frac{dr}{dt} = \frac{\dot{r}}{\dot{t}} = -B\sqrt{1 - \frac{B}{B_*}}. \tag{S167}$$

As $B \to 0$ when $r \to 2GM$, the coordinate speed tends to zero as the object reaches the Schwarzschild radius. We thus see the object asymptotically approaching the Schwarzschild radius, but never crossing it.

(b) The tidal forces acting on the object are given by the geodesic deviation equation. We want to calculate the maximal tidal forces, so let S be a vector pointing from one end to the other. We assume that the object is initially at rest, pointing toward the black hole, and has a length of $2\,\text{m}$, so $S_* = (0, 2\,\text{m}, 0, 0)$. We let the object fall along a geodesic $x(\tau)$ with tangent vector $T = \frac{d}{d\tau}x(\tau)$. Then the geodesic deviation equation is given by

$$\nabla_T \nabla_T S = R(T, S)T. \tag{S168}$$

To calculate the second covariant derivative along the geodesic on the LHS it is easier to go to a parallel transported frame along the geodesic. In particular, we choose an orthonormal frame (tetrad) (e_0, e_1, e_2, e_3) of $T_{x(\tau)}M$ parallel along $x(\tau)$ s.t. $e_0 = T$ and $g(e_i, e_j) = \eta_{ij}$. We also introduce the co-frame $(\theta^0, \theta^1, \theta^2, \theta^3)$ to this basis, such that $\theta^i(e_j) = \delta^i_j$. As the e_i are parallel along $x(\tau)$ we have $\nabla_T e_i = \nabla_{e_0} e_i = 0$, and thus writing $S = S^i e_i$ in this basis the LHS reduces to

$$\nabla_T \nabla_T (S^i e_i) = T^2 S^i e_i = \left(\frac{d^2}{d\tau^2} S^i \right) e_i. \tag{S169}$$

Using the definition of the curvature forms $\Omega^i{}_j$ we can rewrite the RHS as

$$R(T, S)T = R(e_0, S)e_0 = \Omega^i{}_0(e_0, S)e_i. \tag{S170}$$

In this basis the geodesic deviation equation thus reduces to

$$\frac{d^2}{d\tau^2} S^i = \Omega^i{}_0(e_0, S) = S^j \Omega^i{}_0(e_0, e_j). \tag{S171}$$

Explicitly a choice of such an orthonormal frame is given by

$$
e_0 = \frac{\sqrt{B_*}}{B}\partial_t - \sqrt{B_* - B}\partial_r, \quad e_2 = \frac{1}{r}\partial_\theta,
$$

$$
e_1 = -\frac{\sqrt{B_* - B}}{B}\partial_t + \sqrt{B_*}\partial_r, \quad e_3 = \frac{1}{r\sin\theta}\partial_\phi, \tag{S172a}
$$

$$
\theta^0 = \sqrt{B_*}\,\mathrm{d}t + \frac{\sqrt{B_* - B}}{B}\mathrm{d}r, \quad \theta^2 = r\mathrm{d}\theta,
$$

$$
\theta^1 = \sqrt{B_* - B}\,\mathrm{d}t + \frac{\sqrt{B_*}}{B}\mathrm{d}r, \quad \theta^3 = r\sin\theta\,\mathrm{d}\phi. \tag{S172b}
$$

We can also write the Schwarzschild coordinate frame in this new orthonormal frame

$$
\mathrm{d}t = \frac{\sqrt{B_*}}{B}\theta^0 - \frac{\sqrt{B_* - B}}{B}\theta^1, \quad \mathrm{d}\theta = \frac{1}{r}\theta^2,
$$

$$
\mathrm{d}r = -\sqrt{B_* - B}\,\theta^0 + \sqrt{B_*}\,\theta^1, \quad \mathrm{d}\phi = \frac{1}{r\sin\theta}\theta^3. \tag{S173}
$$

We now use Cartan's structure equations to calculate the connection and curvature forms. In our case of an orthonormal frame they read

$$
\omega_{ij} = -\omega_{ji}, \tag{S174a}
$$

$$
\mathrm{d}\theta^i = -\omega^i{}_j \wedge \theta^j, \tag{S174b}
$$

$$
\Omega^i{}_j = \mathrm{d}\omega^i{}_j + \omega^i{}_k \wedge \omega^k{}_j. \tag{S174c}
$$

First we need to calculate the exterior derivatives of the co-frame. Remember that the exterior derivative of a one-form can be calculated as $\mathrm{d}(\omega_i \mathrm{d}x^i) = \partial_j \omega_i\, \mathrm{d}x^j \wedge \mathrm{d}x^i$. We thus find

$$
\mathrm{d}\theta^0 = 0,
$$

$$
\mathrm{d}\theta^1 = \frac{B'}{2\sqrt{B_* - B}}\theta^0 \wedge \theta^1,
$$

$$
\mathrm{d}\theta^2 = -\frac{\sqrt{B_* - B}}{r}\theta^0 \wedge \theta^2 + \frac{\sqrt{B_*}}{r}\theta^1 \wedge \theta^2,
$$

$$
\mathrm{d}\theta^3 = -\frac{\sqrt{B_* - B}}{r}\theta^0 \wedge \theta^3 + \frac{\sqrt{B_*}}{r}\theta^1 \wedge \theta^3 + \frac{\cot\theta}{r}\theta^2 \wedge \theta^3. \tag{S175}
$$

From this we can easily read of the connection forms via the first and second Cartan structure equation. The result is

$$\omega^0_{\ 1} = \omega^1_{\ 0} = \frac{B'}{2\sqrt{B_* - B}}\ \theta^1,$$

$$\omega^0_{\ 2} = \omega^2_{\ 0} = -\frac{\sqrt{B_* - B}}{r}\ \theta^2,$$

$$\omega^0_{\ 3} = \omega^3_{\ 0} = -\frac{\sqrt{B_* - B}}{r}\ \theta^3,$$

$$\omega^1_{\ 2} = -\omega^2_{\ 1} = -\frac{\sqrt{B_*}}{r}\ \theta^2,$$

$$\omega^1_{\ 3} = -\omega^3_{\ 1} = -\frac{\sqrt{B_*}}{r}\ \theta^3,$$

$$\omega^2_{\ 3} = -\omega^3_{\ 2} = -\frac{\cot\theta}{r}\ \theta^3. \tag{S176}$$

The curvature forms can now easily be calculated with the third Cartan structure equation. Relevant in the geodesic deviation equation are only the curvature forms $\Omega^i_{\ 0}$, and they read

$$\Omega^1_{\ 0} = \Omega^0_{\ 1} = -\frac{B''}{2}\ \theta^0 \wedge \theta^1,$$

$$\Omega^2_{\ 0} = \Omega^0_{\ 2} = -\frac{B'}{2r}\ \theta^0 \wedge \theta^2,$$

$$\Omega^3_{\ 0} = \Omega^0_{\ 3} = -\frac{B'}{2r}\ \theta^0 \wedge \theta^3. \tag{S177}$$

We can now write down the components of the geodesic deviation equation S171:

$$\frac{d^2}{d\tau^2}S^0 = 0, \tag{S178a}$$

$$\frac{d^2}{d\tau^2}S^1 = -\frac{B''}{2}S^1, \tag{S178b}$$

$$\frac{d^2}{d\tau^2}S^2 = -\frac{B'}{2r}S^2, \tag{S178c}$$

$$\frac{d^2}{d\tau^2}S^3 = -\frac{B'}{2r}S^3, \tag{S178d}$$

or inserting $B = 1 - r_s/r$

$$\frac{d^2}{d\tau^2} S^0 = 0 \,, \tag{S179a}$$

$$\frac{d^2}{d\tau^2} S^1 = \frac{r_s}{r^3} S^1 \,, \tag{S179b}$$

$$\frac{d^2}{d\tau^2} S^2 = -\frac{r_s}{2r^3} S^2 \,, \tag{S179c}$$

$$\frac{d^2}{d\tau^2} S^3 = -\frac{r_s}{2r^3} S^3 \,. \tag{S179d}$$

We immediately see that the object will be stretched radially and squeezed in angular direction. Note also that the tidal forces do not depend on our initial conditions B_* and as such do not depend on the velocity of the object. They are a purely geometrical effect depending only on the position in spacetime. We thus assume the object to be at rest and oriented feet down toward the black hole at position $r = r_*$, with length $S_*^1 = 2$ m. So for the tidal force in radial direction we get

$$\frac{d^2}{d\tau^2} S^1 = \frac{r_s}{r^3} S_*^1. \tag{S180}$$

We assume the breaking strength of an object of length 2m to be 20g, so that it will break apart at around $470 r_s$.

(c) Interestingly, we see that the tidal forces at the Schwarzschild radius decrease with the mass of the black hole, i.e.,

$$\frac{d^2}{d\tau^2} S^1 = \frac{r_s}{r_s^3} S^1 = \frac{S^1}{r_s^2} = \frac{S^1}{4G^2M^2}. \tag{S181}$$

If we had chosen a supermassive black hole like the one in the center of our galaxy, the object would have had no problem to cross the Schwarzschild radius and would feel the tidal forces only later.

Index

© The Editor(s) (if applicable) and The Author(s), under exclusive license to Springer
Nature Switzerland AG 2022
P. Jetzer, *Applications of General Relativity*, UNITEXT for Physics,
https://doi.org/10.1007/978-3-030-95718-6

Printed in the United States
by Baker & Taylor Publisher Services